AF842184

Sitzungsberichte der Heidelberger Akademie der Wissenschaften
Mathematisch-naturwissenschaftliche Klasse
Jahrgang 1989, 5. Abhandlung

R. Bayer P. Schlosser G. Bönisch H. Rupp
F. Zaucker G. Zimmek

# Performance and Blank Components of a Mass Spectrometric System for Routine Measurement of Helium Isotopes and Tritium by the $^{3}$He Ingrowth Method

With 22 Figures

*Vorgelegt in der Sitzung vom 1. Juli 1989*
*von Otto Haxel*

Springer-Verlag
Berlin Heidelberg New York
London Paris Tokyo Hong Kong

Reinhold Bayer Peter Schlosser Gerhard Bönisch
Hans Rupp Fritz Zaucker Gerhard Zimmek

Institut für Umweltphysik
Universität Heidelberg
Im Neuenheimer Feld 366
6900 Heidelberg, FRG

**ISBN-13: 978-3-540-51710-8 e-ISBN-13: 978-3-642-48373-8**
**DOI: 10.1007/978-3-642-48373-8**

LC card number: 12-8903
This work is subject to copyright. All rights are reserved, whether the whole or part of the material is concerned, specifically the rights of translation, reprinting, reuse of illustrations, recitation, broadcasting, reproduction on microfilms or in other ways, and storage in data banks. Duplication of this publication or parts thereof is only permitted under the provisions of the German Copyright Law of September 9, 1965, in its version of June 24, 1985, and a copyright fee must always be paid. Violations fall under the prosecution act of the German Copyright Law.

© Springer-Verlag Berlin Heidelberg 1989

The use of registered names, trademarks, etc. in this publication does not imply, even in the absence of a specific statement, that such names are exempt from the relevant protective laws and regulations and therefore free for general use.
Typesetting: K+V Fotosatz GmbH, Beerfelden

2125/3140-543210

## Contents

## Abstract

The design and performance of a mass spectrometric system for measurement of helium isotopes and tritium by the $^3He$ ingrowth method is described. The system consists of a commercial helium isotope mass spectrometer and units for quantitative extraction of the dissolved gases from water samples, purification of the extracted gases and helium/neon separation by means of a cryogenic cold trap filled with activated charcoal. The system is computer controlled and capable of fully automated measurement of helium samples. A routine helium isotope measurement ($^3He/^4He$ ratio and $^4He$ concentration) takes about 15 minutes, a tritium measurement takes about 30 minutes. Calibration of the measurements is done using defined aliquots of an air standard which are treated in the same way as a sample. Measurement precision of 40 g water samples is about $\pm 0.2\%$ for $^3He/^4He$ ratio measurement and about $\pm 0.3\%$ for measurement of absolute $^4He$ concentration (accuracy: $\pm 0.5\%$). The $^3He$ detection limit of the system is about $1 \cdot 10^{-16}$ ccSTP corresponding to a tritium detection limit of 0.003 TU using a 500 g water sample stored for 6 months for $^3He$ ingrowth. The blank components of the system and the contribution of blank corrections to the measurement precision are discussed. An intercalibration of the system with a low-level $\beta$-counting system was performed and within the errors of the single techniques no significant deviation was found between the two data sets. A tritium profile from the Weddell Sea, Antarctica, is presented to demonstrate the performance of the system.

## 1 Introduction

Tritium and helium are widely used as tracers in studies of ventilation and circulation of natural water systems (oceans, groundwater; e.g. BEGEMANN and LIBBY, 1957; MÜNNICH and ROETHER, 1967; CLARKE et al., 1969; ATAKAN et al., 1974; JENKINS and CLARKE, 1976; TORGERSEN et al., 1979; SCHLOSSER, 1986; SCHLOSSER et al., 1988). Tritium usually is measured by low-level $\beta$-counting with detection limits on the order of 0.05 to 0.08 TU (ÖSTLUND and BRESCHER, 1982; WEISS et al., 1976; 1 TU means a tritium to hydrogen ratio of $10^{-18}$). Typical $^3He$ signals are on the order of some to some hundred per cent of the $^3He$ concentration of water in solubility equilibrium with the atmosphere. $^3He$ is measured using special helium isotope mass spectrometers with a precision of the $^3He/^4He$ ratio measurement on the order of about $\pm 0.2\%$. This precision allows resolution of the interesting $^3He$ signals observed in the hydrosphere.

Present day tritium levels in the surface waters of the oceans have values of about 3 to 4 TU in the northern hemipshere and about 0.2 TU in the southern hemisphere. Therefore, the resolution (tritium concentration of the water sample divided by the tritium detection limit) of the conventional techniques is not sufficient for delineation of the fine structure of the tritium distribution in the ocean, especially in the southern hemisphere. The high precision achieved in helium isotope measurement allows to measure tritium via $^3He$ ingrowth with a much higher precision and lower detection limit compared to counting techniques (CLARKE et al., 1976). Using a specially developed helium isotope mass spectrometer JENKINS et al. (1985) demonstrated that tritium measurements are possible with a precision of about $\pm 1\%$ (low-level counting technique: $\pm 2.5\%$ to $\pm 5\%$) and a detection limit of about 0.003 TU for oceanic water samples. As oceanic applications usually require data sets on the order of some hundred samples it is necessary to develop the tritium measurement via $^3He$ ingrowth to a routine procedure with a high capacity for efficient use in future studies of the ocean circulation. The purpose of this paper is to demonstrate that this goal can be achieved using a commercial helium isotope mass spectrometer and special procedures for sample preparation, storage for $^3He$ ingrowth, and $^3He$ transfer from the ingrowth containers into the mass spectrometer. Additionally, the parameters and blanks in routine operation of the system for both $^3He$ and tritium measurement are discussed.

## 2 General Procedure

The measurement of helium isotopes includes three steps. The first step is to collect water samples in the field and to store them in appropriate containers for shore based measurement. The second step is quantitative gas extraction from the water samples and storage of the extracted gases in special glass ampoules before they are measured mass spectrometrically in the final step.

In oceanographic applications, samples are drawn from standard Niskin water samplers and stored in pinched-off copper tubes for measurement of helium isotopes and (optionally) tritium by the $^3He$ ingrowth method. Samples for low-level tritium analysis are stored in 1 l glass bottles. For $^3He$ analysis, samples are degassed in a special vacuum extraction system. The extracted gases are transferred into a glass ampoule using water vapor as carrier gas. For optional tritium analysis, the remaining water is sealed off in a glass bulb by flame sealing and stored in freezer for $^3He$ ingrowth. For low-level tritium measurement the procedure is similar. Approximately 500 $cm^3$ of water is transferred into an evacuated 1 liter glass bulb. The glass bulb is pretreated prior to gas extraction to reduce the blank caused by diffusion of helium dissolved in the glass into the water sample. The water sample is degassed by shaking the glass bulb, the gases are transported out of the bulb by water vapor and pumped away. After gas extraction, the glass bulb is flame sealed and stored for typically 4 to 6 months to allow $^3He$ ingrowth from tritium decay.

The sample containers are fitted to the inlet system of the mass spectrometer for measurement. Here the water vapor is removed in a cold trap cooled by liquid nitrogen and all gases except helium and neon are adsorbed on a charcoal trap held at liquid nitrogen temperature. Helium and neon are separated by a cryogenic cold trap. After helium/neon separation the helium fraction is admitted to a dual collector mass spectrometer and $^3He$ and $^4He$ are measured simultaneously applying the peak height method.

## 3 System Configuration

### 3.1 Gas Extraction Units

Gas extraction from the water samples for helium isotope measurement is done in an all metal vacuum extraction line (Fig. 1). Samples contained in copper tubes (CT, volume about 40 $cm^3$) sealed by stainless steel pinch-off clamps (CL; Fig. 1a) are connected to the extraction system via O-ring fitting (OF). A bulb for expansion of the water (EB) and a glass ampoule (GA) containing activated charcoal are fitted to the system in the same way. In the case of additional tritium measurement the expansion bulb is replaced by a special bulb (volume about 200 $cm^3$) used for storage of the water for $^3He$ ingrowth. These bulbs are equip-

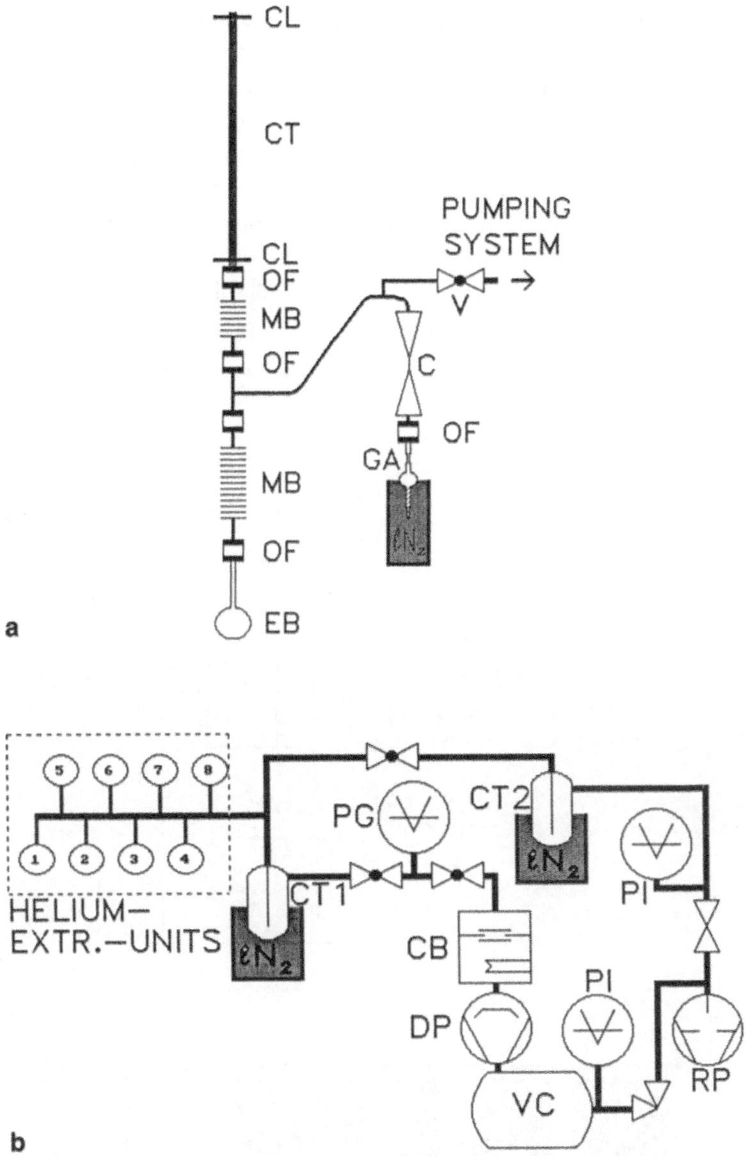


**Fig. 1. a** $^3He$ extraction unit. *CT* copper tube, *CL* stainless steel pinch-off clamps, *OF* O-ring fitting, *MB* metall bellows, *EB* expansion bulb, *C* capillary, *GA* glass ampoule, *V* all-metal valve. **b** $^3He$ extraction system consisting of the vacuum part and 8 extraction units (see **Fig. 1a**). *CT1*, *CT2* cold traps, *PG* Penning cell, *PI* Pirani cell, *CB* cold baffle, *DP* diffusion pump, *RP* rotary pumpe, *VC* vacuum container (provides prevacuum for the diffusion pump)

ped with a glass capillary for flame sealing. Both ampoule and bulb are made of low helium permeability glass (AR-glass; Glaswerke Schott, Mainz). For adjustion of the sample containers and the expansion bulbs flexible metal bellows (MB) are mounted into the system. The water vapor flow is controlled by a capillary (C) in the line connecting the expansion bulb with the glass ampoule. Each single extraction unit is connected to the vacuum line by a metal valve (V). The extraction system consists of the vacuum part and eight extraction units allowing the

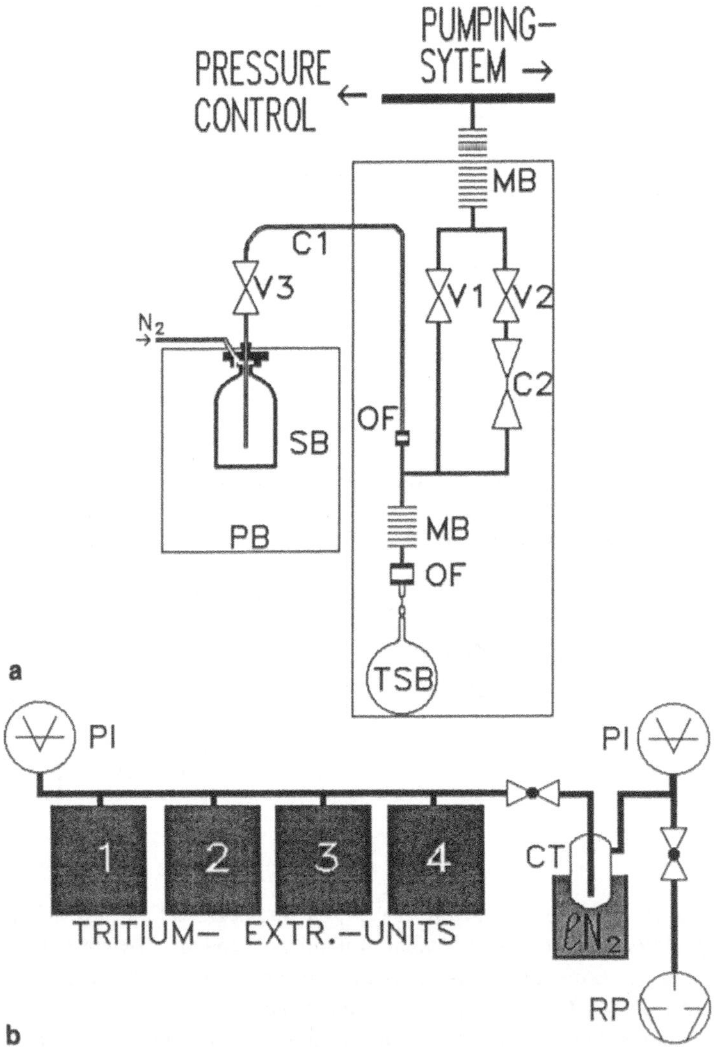


**Fig. 2. a** Tritium extraction unit. *PB* plastic box, *SB* sample bottle, *V1, V2, V3* valves, *C1, C2* capillaries, *OF* O-ring fittings, *MB* metal bellow, *TSB* tritium storage bulb. **b** Tritium extraction system consisting of the vacuum part and 4 extraction units (**Fig. 2a**). *PI* Pirani cell, *CT* cold trap, *RP* rotary pump

parallel processing of eight samples (Fig. 1b). To achieve pressures in the range of $10^{-6}$ mbar a combination of a rotary pump (RP) and a mercury diffusion pump (DP) is used. Pressure is controlled by two Pirani sensors (PI) and one Penning cell (PG). The system can be pumped initially by the rotary pump. High vacuum is established by pumping with the diffusion pump. The pumps are protected against water vapor by cold traps which are held at the temperature of liquid nitrogen. The vacuum container (VC) provides a pressure $\leq 10$ mbar needed for the diffusion pump. The extraction units are bakeable to a temperature of about 100 °C.

Gas extraction from large volume tritium samples (500 $cm^3$) is done in a similar way as helium extraction. The major difference is the fact that for the large tritium samples the extracted gases simply can be pumped away leading to a less complex vacuum system. The sample bottle (SB) is connected via PTFE sealing to a metal seat in which a capillary (C1) and a valve (V3) are fitted for water transfer to the tritium storage bulb (TSB; Fig. 2a). Water transfer is done under nitrogen atmosphere to avoid contamination of the sample by water vapor exchange. For this purpose the sample bottle is handled in a plastic box (PB) flushed with dry nitrogen. The system is pumped via valve V1 to achieve the required pressure of about $10^{-3}$ mbar while V2 is used to pump away the gases during gas extraction. A capillary (C2) restricts the water vapor flow. The tritium sample preparation system consists of 4 gas extraction units and a vacuum part. Vacuum is produced by a rotary pump (RP) which is protected against water vapor by the cold trap CT held at liquid nitrogen temperature. Pressure is controlled by 2 Pirani sensors (PI).

### 3.2 Inlet System

The inlet system consists of the sample inlet, the sample purification and the helium/neon separation parts (Figs. 3 to 5). In the sample inlet part the gas is released by opening the breakseal of the glass ampoules (GA) or tritium storage bulbs (TSB) which are connected to the all metal system via O-ring fittings. The gas is admitted to the sample purification part via a capillary (mounted between the fitting for connection of the sample containers and inlet valve; not pointed out in Fig. 3) to reduce water vapor flow. Additionally, air standard aliquots of different size for standardization of the measurements are provided in the inlet part. The aliquots can be taken by an automated pipette (PA). Rough vacuum can be produced by a rotary pump (RP). In the sample purification part (Fig. 4) water vapor is removed by cold trap CT cooled by liquid nitrogen. All gases except helium and neon are adsorbed on activated charcoal held at liquid nitrogen temperature (ACT). The sample purification part is pumped by a combination of a rotary pump (RP) and a diffusion pump (DP). Pressure is controlled by a Penning gauge (PG) and by Pirani sensors (PI).

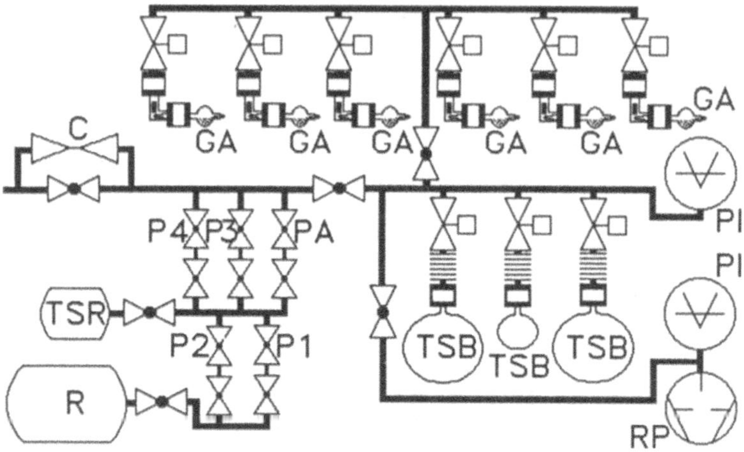


**Fig. 3.** Sample inlet. *GA* glass ampoule, *C* capillary, *PI* Pirani cell, *R* air standard reservoir, *P1–P4, PA* air standard pipettes, *TSB* tritium storage bulb, *RP* rotary pump, *TSR* tritium standard reservoir

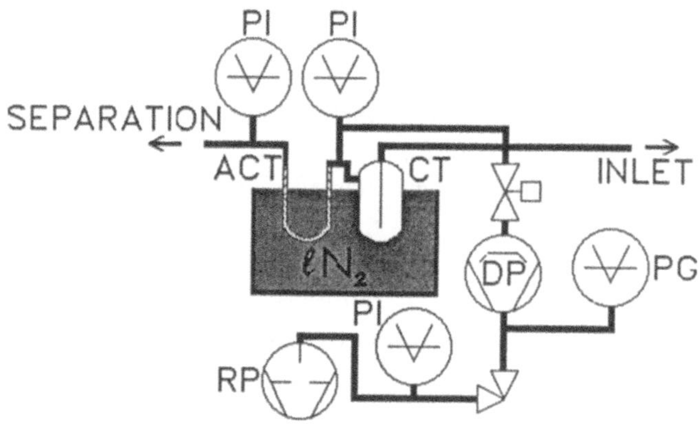


**Fig. 4.** Sample purification. *PI* Pirani cell, *ACT* cold trap filled with activated charcoal, *CT* cold trap, *DP* diffusion pump, *RP* rotary pump, *PG* Penning cell

For helium/neon separation a cryogenic cold trap with activated charcoal as adsorber is used (REYNOLDS et al., 1978). The cold trap is cooled by a helium compressor. The cold head is mounted in a vacuum chamber pumped by a rotary pump (RP 1) to supress heat exchange. Vacuum is provided by a combination of a rotary pump (RP 2), a diffusion pump (DP) and a turbomolecular pump (TP) to achieve clean ultra-high vacuum. Pressure control is done by Pirani (PI) and Bayard-Alpert (BA) sensors. A getter pump (GP) operated at room temperature is mounted in the ultra high vacuum line to reduce the hydrogen level. The samples are transferred to the mass spectrometer (MM 3000) by volume partitioning.

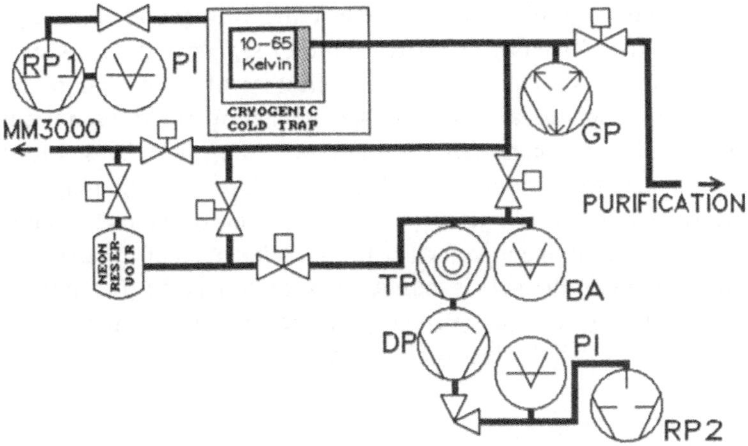


**Fig. 5.** Helium/neon separation. *RP1, RP2* rotary pumps, *PI* Pirani cell, GP getter pump, *TP* turbomolecular pump, *BA* Bayard-Alpert cell, *DP* diffusion pump. The neon reservoir is used to store the neon fraction of a sample if neon is measured in parallel to helium isotopes

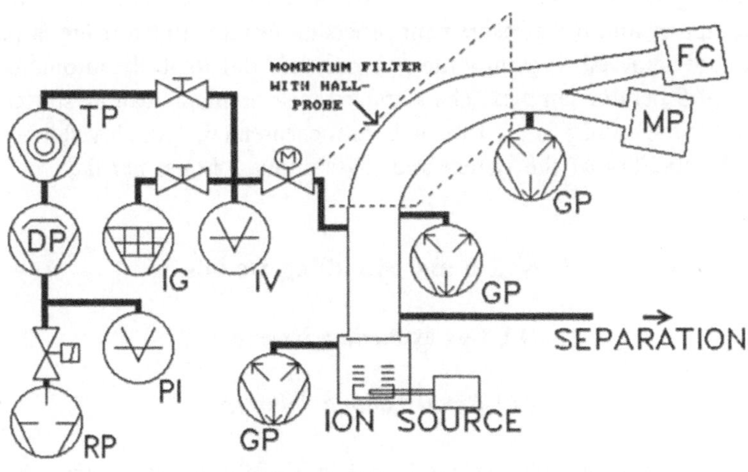


**Fig. 6.** Mass spectrometer. *FC* Faraday cup, *MP* electron multiplier, *GP* getter pump, *IV* Ionivac cell, *IG* ion getter pump, *TP* turbomolecular pump, *DP* oil diffusion pump, *PI* Pirani cell, *RP* rotary pump

### 3.3 Mass Spectrometer

The MM 3000 is a statically operated, dual collecting, dedicated helium isotope mass spectrometer (Fig. 6; LYNCH and KAY, 1980). The $^{4}He$ ion beam is measured with a Faraday Cup (FC). $^{3}He$ is detected simultaneously using a Johnston-MM 1 electron multiplier (MP) operated in the ion counting mode.

Three getter pumps (GP) are used to reduce the hydrogen level during measurement. A valve (M) driven by an electrical motor separates the mass spectrometer from the pumping system. To achieve a pressure of $10^{-9}$ mbar or better, two alternative pumping lines are available. A combination of a turbomolecular pump (TP), an oil diffusion pump (DP) and a rotary pump (RP) is used for pumping of the mass spectrometer after ventilation and for pumping away the samples after measurement. An ion getter pump (IG) serves to maintain vacuum during times between measurement periods or for pumping of smaller gas loads (e.g. during tritium analysis). Pressure is monitored by an Ion gauge (IV) in the ultra high vacuum part and by a Pirani (PI) gauge in the fine vacuum part of the system. The whole system is bakeable to about 350 °C.

### 3.4 Computer Control

Sample inlet and measurement is fully automated. Setting and control of parameters as well as data collection and evaluation is done by an HP 85A computer. The high vacuum valves in the inlet system are pneumatically driven and controlled by the computer. All pressure sensors of the inlet system are monitored by the computer and the measurement procedure is interrupted if preset pressure levels are not achieved. Presently the system is laid out for fully automated measurement of 9 helium samples. The standard measurements can be started some hours before the system is used for helium measurements which yields information on the stability of the system and saves about 2 hours per day.

## 4 System and Handling Details

### 4.1 Gas extraction Systems

#### *4.1.1 Small Volume Samples*

Prior to operation, the system is baked for 1 hour at a temperature of $\approx 100\,°C$ and pumped to a pressure of $\approx 10^{-5}$ mbar. Cooling of cold trap CT 1 with liquid nitrogen reduces the pressure to about $2 \cdot 10^{-6}$ mbar. The sample is transferred from the copper tube to the glass bulb by removal of the lower pinch-off clamp, pressing of the copper seal and heating of the copper tube to ensure that all water is transferred into the expansion bulb. During expansion of the water into vacuum it is degassed to about 80%. After completion of the water transfer the ampoule is cooled by liquid nitrogen to establish a strong water vapor flow from the head space of the bulb over the capillary into the ampoule. The bulb is shaken by a motor driven excenter to increase the transfer velocity of dissolved gases into the head space. Water vapour is frozen out on the glass walls of the ampoule while permanent gases are adsorbed on the activated charcoal. Gases which can not be adsorbed on the charcoal quantitatively (helium and neon) are trapped inside the ampoule by the strong water vapour flow through the glass capillary ($^3$He would have to diffuse against a high counter advection of about 50 m/s under given conditions). This process continues for about 30 minutes allowing all the helium in dead volumes of the extraction unit (e.g. the copper tube) to diffuse into the water vapour flow as well as to flush the bulb's head space at a sufficient rate. The sample is sealed off by flame sealing of the ampoule. For optional tritium analysis, the water in the storage bulb is further pumped for a few minutes to ensure complete degassing. The bulb is sealed off and stored in a freezer for $^3$He ingrowth.

The efficiency of the helium extraction has been determined by threefold extraction of samples resulting in constant $^4$He values of $(2.2 \pm 0.2) \cdot 10^{-8}$ ccSTP for the second and third extraction (about 1% of the helium extracted from the sample in the first step). This value is expected from estimation of helium diffusion through viton O-rings (we use seven O-rings in each extraction unit) and from helium released from the glass during the flame sealing (see below). On the basis of the constant $^4$He values obtained from the second and third extraction of a water sample, we conclude that the gas extraction efficiency is about 99.9% or better.

### *4.1.2 Large Volume Samples*

The system is pumped to a pressure $< 10^{-2}$ mbar and the cold trap is cooled by liquid nitrogen. The sample bottle is opened in the plastic box under nitrogen atmosphere and mounted to the metal seat containing the capillary for water transfer. Sample transfer is done by opening valve V 3 (Fig. 2b) while pumping valves V 1 and V 2 are closed. Due to the pressure difference water flows through capillary C 1 into the tritium storage bulb (TSB). The head space of the sample bottle is held at a constant nitrogen pressure to avoid tritium contamination. After 5 minutes, about 500 cm$^3$ of water is transferred. The gases released to the head space of the bulb (on the order of 80% of the helium) during water transfer are removed by pumping via valve V 1. In the final step of the extraction procedure the gases are pumped via capillary C 2 in order to reduce the water vapor flow. Fast transfer of the dissolved helium to the head space is achieved by intense stirring of the water. The extracted helium is removed from the bulb by water vapor diffusion pumping. The water vapor flow is adjusted by the glass capillary in a way that rediffusion of helium into the bulb is suppressed. Water vapor is frozen out in the cold trap while permanent gases are removed by the rotary pump. After 30 minutes the bulb is sealed off by flame sealing.

Extraction efficiency has been estimated from time series, i.e. gas extraction of a set of samples using different extraction times. Helium measurements have been performed directly after gas extraction to reduce blank contributions arising from storage of the samples in the glass bulbs (see below). The results of these measurements yield residual helium levels in the extracted water sample of less than $2 \cdot 10^{-10}$ ccSTP $^4$He for extraction times of more than 25 minutes. In this way, the extraction efficiency is estimated to be higher than 99.999% for a typical sample size of about 500 cm$^3$ water.

## 4.2 Inlet System

### *4.2.1 Sample Inlet*

The sample is attached to the inlet system by a viton O-ring fitting. Tritium samples are thawed and stirred to ensure that tritogenic $^3$He is transferred to the head space of the bulb. The inlet system is evacuated by the diffusion pump to a pressure of better then $10^{-5}$ mbar and the samples is transferred by opening the breakseal of the glass bulb. Gases present in the head space are transferred to the purification part by water vapour flow via a capillary. As described above, the capillary acts as a barrier against back diffusion of helium into the glass bulbs (or ampoules in the case of $^3$He measurement). Because of the low solubility of helium in seawater (less than 1%), more than 99% of the helium is in the head space of the sample container. For quantitative transfer to the inlet system, the head space must be flushed at a sufficient rate. This rate was determined as follows: an all metal bulb (volume about 400 cm$^3$) filled with 200 cm$^3$ of water and mounted via an all metal fitting to the inlet part (to prevent helium diffusion into the system via O-rings) was degassed on-line to about 80% and the extracted gases were removed by the rotary pump. After a time sufficiently long to reach helium solubility equilibrium (i.e. more than 99% of the residual helium was transferred to the head space), the bulb was opened for a few seconds to allow transfer of a part of the 'sample'. Then purification, helium/neon separation and measurement were done routinely. This process was repeated until helium concentrations were below the detection limit. An exponential curve was fitted to the results (Fig. 7). Above 100 seconds accumulated inlet time the slope of the curve indicates quantitative transfer. The e-folding time of the curve can be used to calculate the adequate transfer times for helium from ampoules and from tritium bulbs into the inlet system. In the case of helium ampoules (1 cm$^3$ water and about 25 cm$^3$ head space) about 15 seconds are needed for quantitative helium transfer into the inlet system. The corresponding time for 40 cm$^3$ tritium samples (Head space about 200 cm$^3$) is 100 seconds, and for large volume tritium samples (500 cm$^3$ for both water and head space) is 250 sec. Quantitative transfer has been checked by twofold extraction of the same water samples. Helium still dissolved in the water after the extraction is only a small fraction of the total helium and can be corrected for.

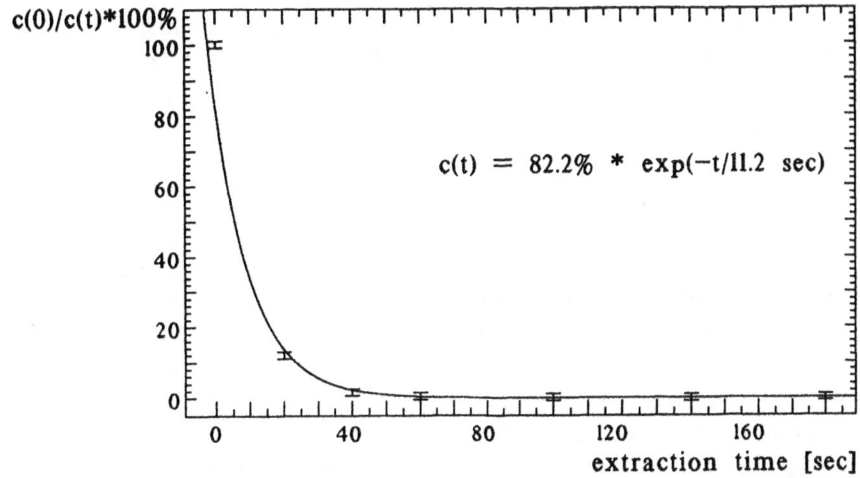


**Fig. 7.** Transfer of helium from the tritium storage bulbs into the inlet system. For explanation see text

### *4.2.2 Air Standards*

For standardization of the measurement, air standards of different size are available. Air is drawn from a 10 l stainless steel reservoir (R). Filling of the container is done under defined conditions and the partial pressure of helium is calculated with an accuracy of about ±0.3%. For $^3He$ and $^4He$ analysis, the primary standard is taken by an automated pipette (PA; Fig. 3). The aliquot volume of $0.3817 \pm 0.0012\ cm^3$ is chosen to provide standards of about $1.8 \cdot 10^{-6}$ ccSTP $^4He$, which is close to helium contents of typical oceanographic samples (40 g water). The accuracy of the helium aliquot is estimated to about ±0.4%. The standards are expanded to the inlet and purification system and processed in the same way as samples (see below). Depending on the daily trend of the standard curve, two or three samples are bracketed by an air standard analysis. For long term calibration of the automated pipette, a similar aliquot volume (P 3; $0.3752 \pm 0.0012\ cm^3$) is used. Checks of the system linearity and mass discrimination (see below) are performed by measurement of one, two, three or four aliquots of $0.1558 \pm 0.008\ cm^3$ of air provided by pipette P 4.

For tritium analyses, air standards which are three orders of magnitude lower than those used for helium measurements are required. Aliquots are drawn by the automated pipette from a tritium standard reservoir (TSR, Fig. 3; volumes: $585.11 \pm 1.75\ cm^3$) filled by expansion of standard volumes P 1 or P 2 ($0.7925 \pm 0.0038\ cm^3$ and $0.1028 \pm 0.0011\ cm^3$, respectively). Single, two- and threefold aliquots are used to calibrate the system over the range of interest. Reduction of the helium concentration of the air reservoir due to continuous drawing of aliquots must be corrected for. The typical size of an aliquot obtained by this technique is $2.5 \cdot 10^{-9}$ ccSTP $^4$He. The corresponding $^3$He size of $3.4 \cdot 10^{-15}$ ccSTP $^3$He is close to the $^3$He production by tritium decay in a 0.1 TU sample of 500 g in 6 months. The helium content of an aliquot is calculated with an accuracy of $\approx \pm 0.7\%$.

### *4.2.3 Sample Purification*

The water vapor introduced by sample or standard inlet is removed by a cold trap, and permanent gases are adsorbed by passing a U-tube filled with activated charcoal ($\approx 10$ g), both held at liquid nitrogen temperature. Pirani sensors at both ends of the U-tube are used to monitor the purification process and to check the samples with respect to oversampling or leakage. After $\approx 5$ minutes, all the gases except helium and the major fraction of neon are adsorbed.

### *4.2.4 Helium/Neon Separation*

A cryogenic cold trap is used to separate helium from neon before it is admitted to the mass spectrometer. Helium/neon separation is a necessary step in order to avoid mass discrimination of helium by neon. Helium/neon separation is done by adsorption of helium and neon on a charcoal trap cooled to 10 K followed by heating the charcoal trap to a temperature which allows desorption of helium. At this temperature the neon should desorb only to a small fraction (some per cent). The cold trap consists of a cold head filled with about 2.5 g of activated charcoal (F 12; Degussa, Frankfurt; active surface $\approx 1200$ m$^2$/g). The individual charcoal grains are glued onto the metal cold head chamber walls using a high thermal conductivity epoxy glue to ensure fast thermal transition. The copper-tellur chamber is mounted directly on the expander head of a helium cryogenic system. A good thermal isolation is provided by a secondary vacuum chamber pumped by a rotary pump allowing fast temperature changes of the cold trap between 10 and 70 K. To avoid any thermal lag by heat radiation, a special shield held at 80 K is mounted around the cold head chamber. To minimize heat transfer from the processing line, a stainless steel bellow (thickness: 0.2 mm) is used for connection of the cold head. Additional copper shields in front of the cold head serve to reduce heat radiation. The temperature of the cold trap can be set by computer control with an accuracy of $\approx \pm 0.15$ K. The cold trap can be cooled from 65 K to 10 K in less than 6 minutes.

The helium desorption curve as a function of the temperature of the cold trap was determined using a separate air standard aliquot ($\approx 1.8 \cdot 10^{-6}$ ccSTP) for each release temperature (5 K intervals between 10 and 90 K). The helium and neon components of these standards were drawn through the purification part into the cryogenic cold trap (10 K) for 8 minutes. After heating the cold trap to the

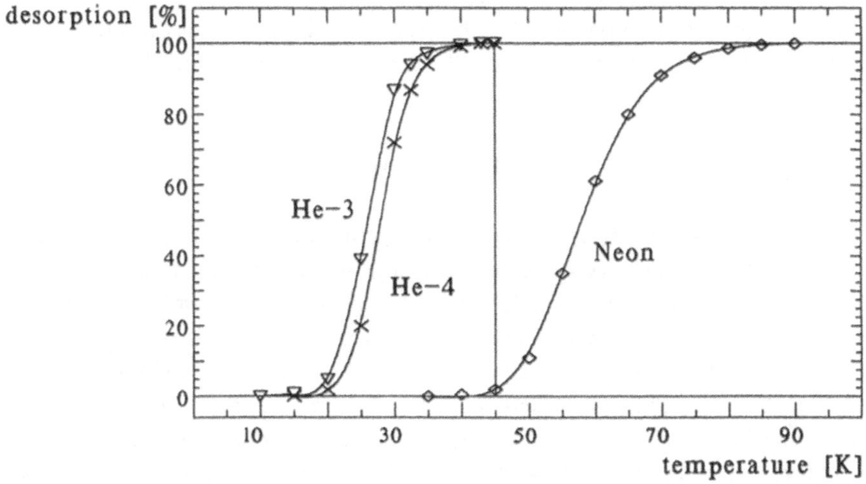


**Fig. 8.** Adsorption of $^{3}He$, $^{4}He$ and Ne on the cryogenic cold trap as a function of temperature

release temperature the gas phase was allowed to reach equilibrium with the mass spectrometer before helium isotope measurement. Due to the volume split between the helium/neon separation system and the mass spectrometer (200 $cm^3$ and 1000 $cm^3$, respectively) about 80% of a sample is transferred into the mass spectrometer. At 10 K both $^{3}He$ and $^{4}He$ are adsorbed to more than 99.9% (Fig. 8). At 45 K helium is desorbed nearly to 100% while neon is almost completely adsorbed on the charcoal. Above 40 K there is no observable helium isotope fractionation. At 50 K about 10% of the neon is desorbed and the $^{4}He$ signal decreases because of the mass discrimination effect in the ion source (not pointed out in Fig. 8). Neon desorption reaches practically 100% at about 87 K. The desorption curves for the neon isotopes $^{20}Ne$ and $^{22}Ne$ are identical within the experimental error. On the basis of the helium and neon desorption curves the cold trap temperatures are set to 10 K for gas transfer from the purification system to the helium/neon separation system and to 45 K for helium desorption and expansion into the mass spectrometer. After gas transfer to the mass spectrometer the cold trap is heated to 65 K to release most of the neon which is pumped away using a turbomolecular pump. Complete regeneration of the cold trap is done by heating it to room temperature after a measurement period of typically 20 hours. Because of the smaller gas load, transfer times and temperatures obtained for helium analysis hold true for tritium measurement, and regeneration of the cold trap is not necessary for a longer time allowing a 24 hour schedule for tritium measurements. The times required for quantitative helium adsorption on the cold trap and for desorption and expansion into the mass spectrometer were determined by variation of the adsorption and desorption times for air standard aliquots of the same size. We obtained e-folding times of 10 seconds for adsorption on the cold trap and about 15 seconds for desorption and gas expansion into the mass spectrometer. The routine time intervals were set to 3 minutes for both processes.

### 4.3 Mass Spectrometer

The MM 3000 is an all-metal, statically operated 90° sector mass spectrometer equipped with a Clarke type electron-impact ion source. The source is operated at a filament current of typically 5 A yielding an electron emission current of about 5 mA regulated by a trap current of 1 mA. The electron energy is adjusted to about 100 eV. The sensitivity of the source for helium is about $2.7 \cdot 10^{-4}$ A/mbar. The instrument is capable of a resolution of about 1500 (10% valley criterium). For routine measurement it is operated at a resolution of $\approx 600$ to achieve flat peak tops. The width of the $^3He$ peak is 1.8 V, and within a range of $\pm 0.9$ V around a given voltage for the center of the peak the ion current decreases by less than 3%. Stability of the acceleration voltage is usually better than $\pm 0.09$ V which leads to a negligible error of $\pm 0.03\%$ if a parabolic shape of the peak top is assumed. Contribution of HD and $H_3^+$ (masses 3.022 and 3.023, respectively) to the mass of $^3He$ (3.016) is small ($< 0.01\%$ of the $^3He$ content of a 40 $cm^3$ seawater sample) and can be corrected for.

The $^4He$ ion beam ($\approx 3 \cdot 10^9$ ions/sec for a sample of $1.8 \cdot 10^{-6}$ ccSTP helium or $\approx 4 \cdot 10^{13}$ atoms $^4He$ in the mass spectrometer) is collected on a Faraday cup yielding a signal of about 4.5 V ($10^{10}\ \Omega$ resistor). The electrometer output is digitized and the integration is done using digital counters. The $^3He$ ions ($\approx 3000$ ions/sec for $2.5 \cdot 10^{-12}$ ccSTP or $5.5 \cdot 10^7$ atoms $^3He$ resulting from $1.8 \cdot 10^{-6}$ ccSTP of helium with atmospheric $^3He/^4He$ ratio) are collected on a focussed mesh electron multiplier (MM 1; Johnston Laboratories) with a gain of $10^6$ at a voltage of 3 kV. The MM 1 is operated in a pulse counting mode with an electronic dead time of $\approx 5$ µsec. A comparison of the $^3He$ signal with the $^4He$ signal yields an efficiency of the $^3He$ detection (ratio of detected ions to ions reaching the first dynode) of $\approx 70\%$. The noise of the counting electronics is typically $\approx 0.001$ cps, the noise of the multiplier without ion beam is $\approx 0.03$ cps. The background of the pumped system ('white noise') has typical values of $\approx 0.10$ cps which most likely is caused by scattered hydrogen ions. The contribution of this background to the $^4He$ signal is negligible. The ion optical transmission (ratio of the detected ions to the ions produced in the ion source) was estimated roughly by comparison of the detected $^4He$ ions with the number of ions removed from the gas phase during measurement of an air standard, and a value of about 20% was obtained.

## 4.4 Measurement of the $^3He/^4He$ Ratio

### *4.4.1 Procedure*

During the expansion of the sample from the cryogenic cold trap into the mass spectrometer (2 min) peak adjustment by scanning both the magnetic (rough adjustion) and the electric field (fine adjustion) is started. After closing the inlet valve and finishing the scans, $^3He$ and $^4He$ are measured simultaneously. One analysis consists of 30 single steps divided into integration on peak (15 sec) followed by integration on background (3 sec) leading to total integration times of 450 sec (peak) and 90 sec (background). The background corrected $^4He$ signals are fitted by a line and the value obtained by extrapolation to time zero is used for further evaluation. The $^3He/^4He$ ratio is obtained by calculating the mean of the 30 background-corrected values. The $^4He$ peak height stability during an analysis of about 10 min is about 0.01% for an aliquot of $1.8 \cdot 10^{-6}$ ccSTP $^4He$. As the error due to ion counting is negligible ($\approx 10^{12}$ ions are detected) the peak height stability is dominated by small variations of the sensitivity. The error of the $^3He/^4He$ ratio measurement is ion counting limited. A signal of $\approx$3000 cps ($2.5 \cdot 10^{-12}$ ccSTP $^3He$) and a background of $\approx$20 cps (scattered hydrogen and $^4He$ ions) result in an error of about $\pm 0.1\%$ (450 sec integration time). This value is in good agreement with the standard deviations obtained for the mean value of the 30 single integrations (typically $\pm 0.15\%$; range: $\pm 0.09\%$ to $\pm 0.25\%$) which shows that systematic errors are small. The cycle time for a single analysis (sample inlet, peak adjustion, integration, evaluation) is about 15 min. Usually 3 samples are measured between air standards which leads to a capacity of 3 samples per hour.

### *4.4.2 Precision and Reproducibility*

Calibration of the measurements is done by fitting a polynomial curve (up to an order of 3) to the air standard measurements obtained during a day (typically 10 to 15 standard measurements). The trends in the air standard measurements reflect small variations of sensitivity (drift may be in excess of about 1% during a day) and sample preparation parameters (mainly temperature fluctuations of the cryogenic helium/neon trap). The scatter around the fit curves (Fig. 9) is a measure of the reproducibility of the measurements and used to check the performance of the system. Typical values of the scatter around the fit curves are $\pm 0.15\%$ for $^3He/^4He$ measurement and $\pm 0.3\%$ for $^4He$ measurement. Together with the uncertainty of the standardization (see above), the accuracy of $^4He$ analysis is $\approx \pm 0.5\%$.

Overall reproducibility (sampling, gas extraction and mass spectrometric measurement) is checked routinely by measuring sets of 4 to 8 samples drawn from

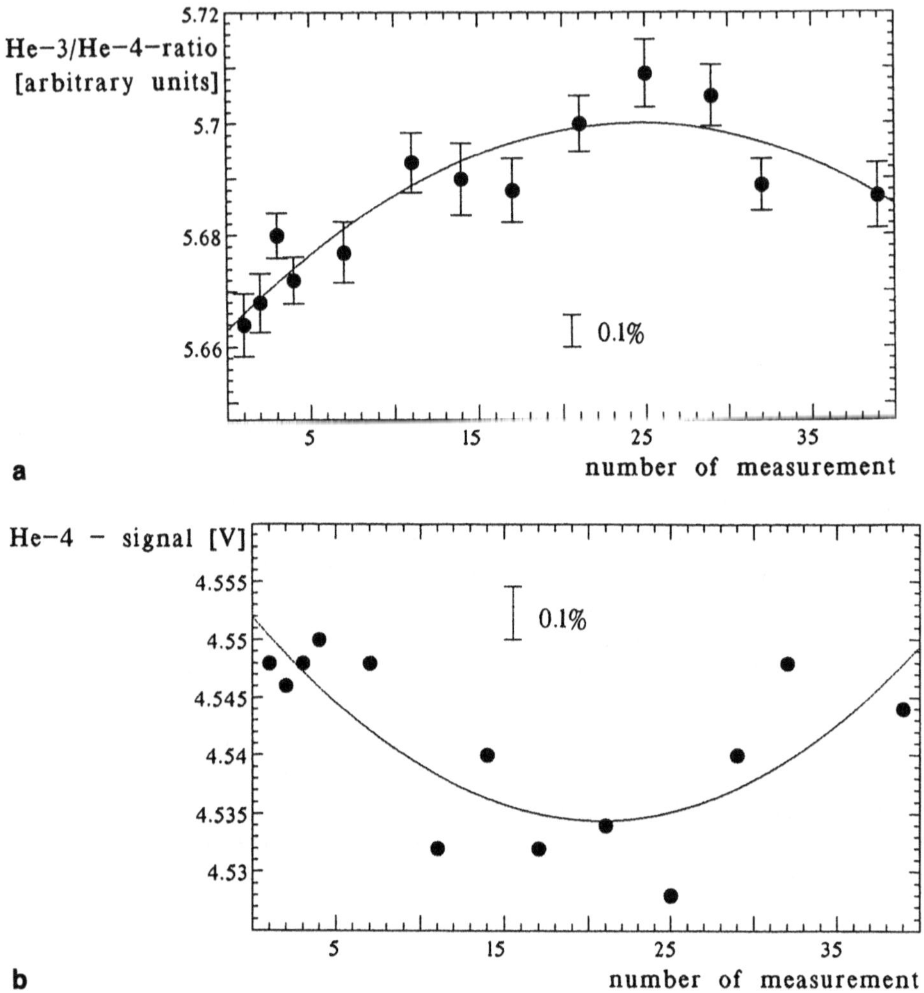


**Fig. 9. a** Typical $^3He/^4He$ results from air standard measurements over one day with 1 $\sigma$ error bars (scatter around the fit curve of $\pm 1.2‰$. **b** same as 9a for $^4He$. The scatter around the fit curve is $\pm 1.0‰$

the same water reservoir in the laboratory. The results of these measurements usually are in good agreement with the errors of the single measurements and the reproducibility of the air standards (Fig. 10). This fact gives confidence that no significant additional errors are introduced during the sample processing steps.

### *4.4.3 Non-linearity Effects*

The mass spectrometer response to changes of the partial pressure of helium in the ion source is non-linear for both $^4He$ and the $^3He/^4He$ ratio. This non-

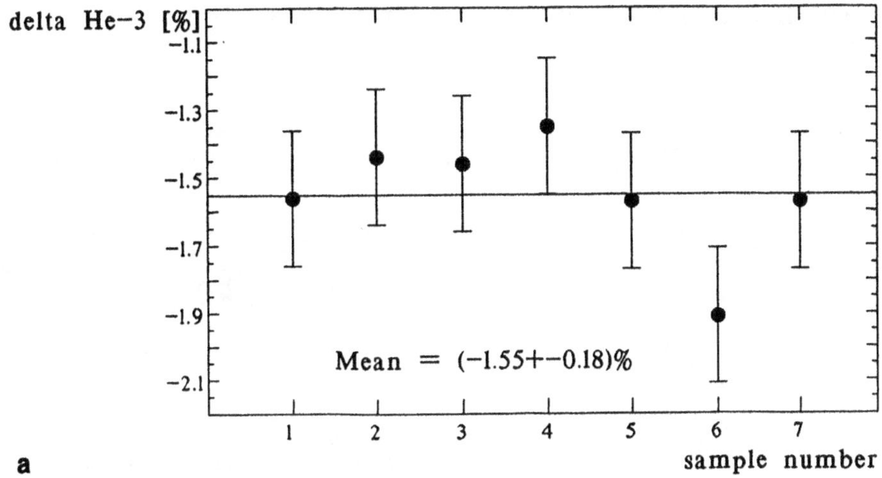


a

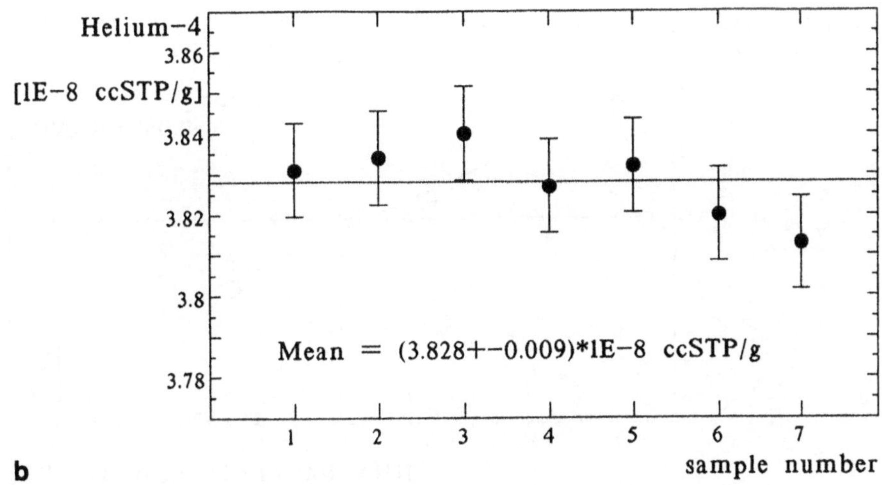


b

**Fig. 10. a** Replicate $^3He$ measurements of a sample set drawn from a water reservoir close to solubility equilibrium in the laboratory. The horizontal line denotes the mean value. **b** same as 10a for $^4He$

linearity is caused by mass discrimination and is dependent on the tuning of the ion source. Therefore, it has to be checked routinely by measurement of a suite of air standards of different size (1 to 4 aliquots of volume P4; see Fig. 3). In general a decreasing sensitivity with increasing sample size for both $^4He$ and the $^3He/^4He$ ratio is found (Fig. 11). Typical values of the sensitivity loss are $-0.03\%$ per 1% sample size for $^4He$ and $-0.07\%$ per 1% sample size for the $^3He/^4He$ ratio. In oceanographic applications with $^4He$ concentrations varying by about $\pm 20\%$ the effects of non-linearity are small and can be corrected for without introducing considerable errors.

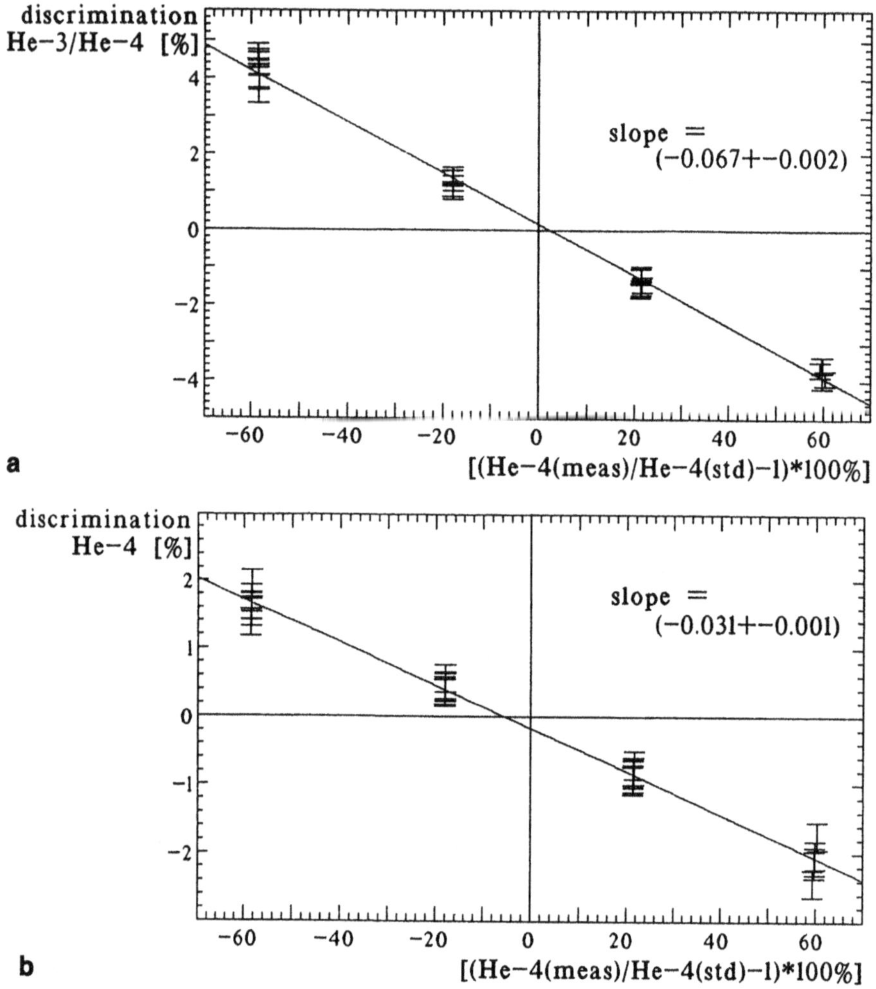


**Fig. 11.** **a** Discrimination of the $^3He/^4He$ measurement as a function of the $^4He$ concentration in the mass spectrometer. The measurements were obtained over a period of 2 weeks. **b** same as 11a for $^4He$

For samples with $^3He/^4He$ ratios below $10^{-7}$ the contribution of the $HD/H_3^+$ peak to the total signal on mass 3.016 ($^3He$) is more important and has to be corrected for. For measurement of samples with $^3He/^4He$ ratios below $10^{-7}$ the peak adjustion is done indirectly by a scan of the $HD/H_3^+$ peak followed by a high voltage switch to the mass of $^3He$ (see also under tritium measurement).

## 4.5 Tritium Measurement

### *4.5.1 Procedure*

Since tritium concentrations are determined via measurement of $^3$He produced by tritium decay in a *degassed* water sample, typical gas loads admitted to the mass spectrometer are 2 to 3 orders of magnitude lower than those for helium measurement. Therefore, adjustion of the $^3$He peak is done by locating the $HD/H_3^+$ peak followed by a change of the electric field to the mass of $^3$He. Due to the low $^3$He signals which are relatively close to the background signal, one analysis consists of 10 single cycles of integration on peak (90 sec) and on background (48 sec). $^4$He is evaluated in the same way as done for helium measurement (see above). $^3$He evaluation is done by calculating the mean of the 10 background-corrected integrations. The total time required for a tritium measurement is 30 min including sample inlet, peak adjustion, integration and evaluation of the signals. For a typical air aliquot size of $2.5 \cdot 10^{-9}$ ccSTP $^4$He (about $5.4 \cdot 10^{10}$ atoms in the mass spectrometer) and $3.5 \cdot 10^{-15}$ ccSTP $^3$He ($7.5 \cdot 10^4$ atoms) and the corresponding count rates of $\approx 4 \cdot 10^6$ ions/sec $^4$He and $\approx 3$ ions/sec $^3$He, measurement precision is limited by ion counting statistics to $\pm 0.002\%$ for $^4$He and to about $\pm 2\%$ for $^3$He, respectively. For $^3$He the value calculated from the precision obtained for the single measurement and the standard reproducibility is close to this estimate. The precision typically achieved for $^4$He again is in excess to $\pm 0.1\%$ due to short term fluctuations in sensitivity.

### *4.5.2 Standardization*

Measurements are calibrated using air standard aliquots that are reduced by volume splitting to a size of $3.5 \cdot 10^{-15}$ ccSTP $^3$He. This corresponds to the $^3$He signal produced by a 500 cm$^3$ water sample with a tritium concentration of 0.1 TU during a storage time of 6 months. The aliquot used for calibration is known with an accuracy of about $\pm 0.7\%$ including uncertainties in the volume of the air standard pipettes and in the partial pressure of the air standard reservoir. During tritium measurement periods, sensitivity of the mass spectrometer for both $^3$He and $^4$He is checked by analysis of air aliquots of different size (usually 1 to 3 aliquots). The signals obtained for analysis of 'zero' aliquots are a measure of the blank of the system. Usually two or three samples are bracketed by measurement of standards or line blanks. An example for a typical analysis period is given in Fig. 12.

Data processing is done separately for $^3$He and $^4$He in identical fashion. Sensitivity is calculated from the slope of a linear fit through both the air standard and the line blank data. The zero intercept of the fit curve gives an averaged determination of the system's operating blank and usually is in good agreement with

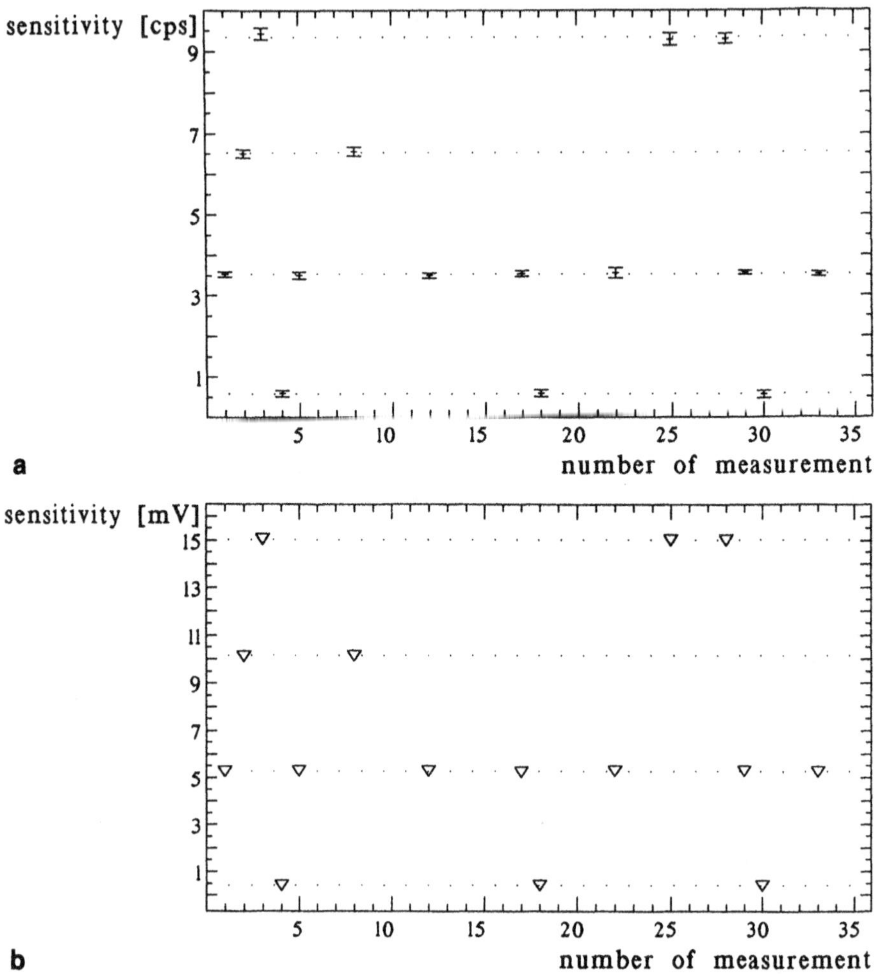


**Fig. 12. a** Typical results of $^{3}He$ standard measurements used for calibration of tritium measurements over one day. In this case we measured one, two and three aliquots of an air standard in addition to line blanks. **b** same as 12a for $^{4}He$

the line blanks measured directly. The scatter around the fit curve is an estimate of the detection limit (smallest amount of helium detectable above background scatter). Given a $^{3}He$ detection limit, the minimum detectable tritium concentration depends on both the time of storage and the amount of degassed water.

### *4.5.3 $^{3}He$ Detection Limit*

Detection limits for both $^{3}He$ and $^{4}He$ obtained during a tritium measurement period are summarized in Fig. 13. The $^{3}He$ detection limit can be lowered

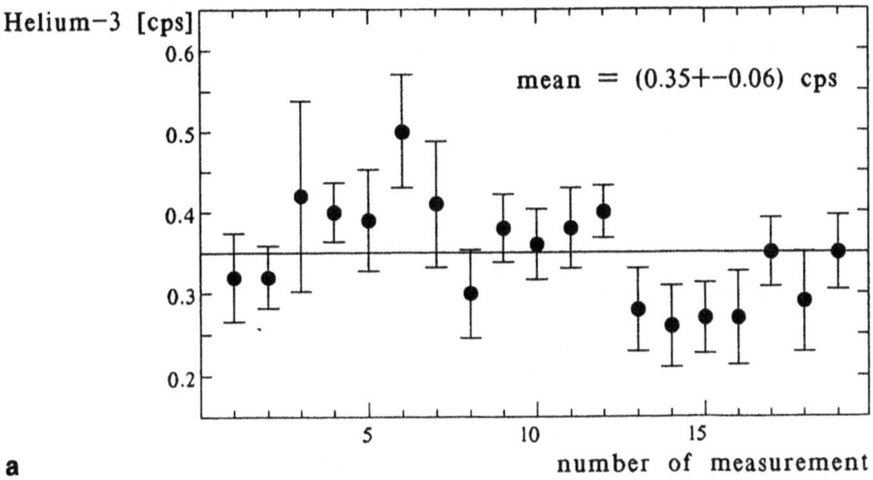


**a**

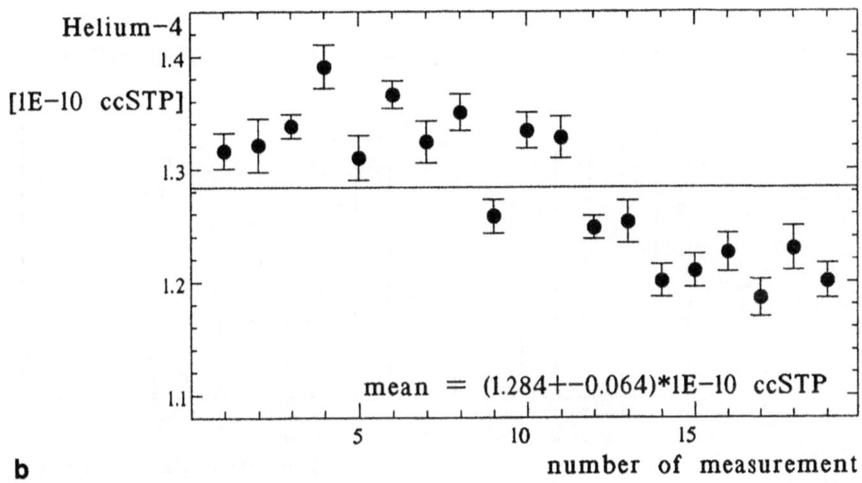


**b**

**Fig. 13 a, b.** Line blank measurements for $^3He$ **a** and $^4He$ **b** over a period of one week. The relatively high error of the mean $^3He$ line blank (0.06 cps) is caused by a long term variation over a week. Usually we obtain values in the range between 0.03 and 0.05 cps

by increasing the integration time, but the ultimate limit is determined by the precision of the background measurement. Typical background values ('white noise' measured at the mass 3 *background* position) of the pumped spectrometer are on the order of 0.1 cps. The noise increases to about 0.15 cps when the spectrometer is closed. This value is not changed if helium with concentrations $<2 \cdot 10^{-8}$ ccSTP is admitted to the mass spectrometer, which shows that for low helium concentrations the 'white noise' is dominated by scattered hydrogen. The typical $^3He$ line blank (defined as the mean $^3He$ signal measured on mass 3.016 in 26 minutes integration, i.e. the time of a tritium analysis, minus the 'white

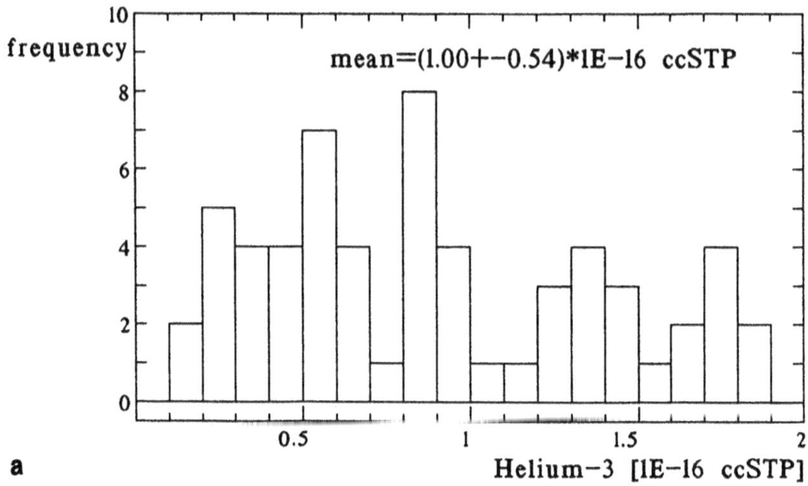


a

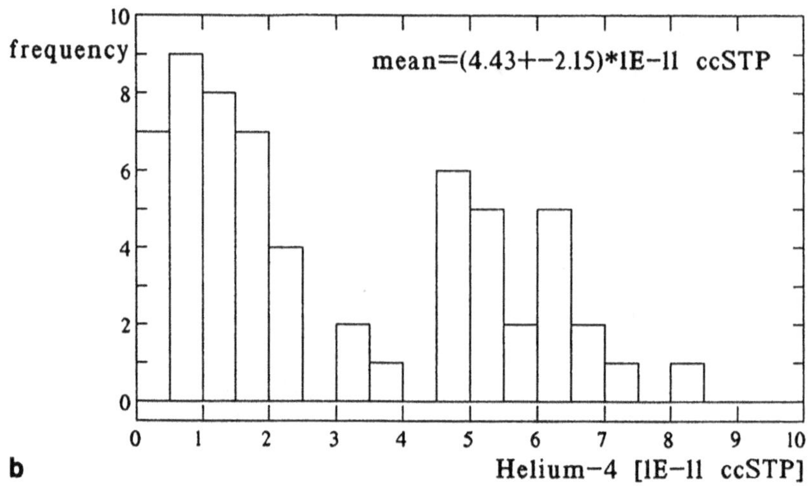


b

**Fig. 14a, b.** Distribution of $^3He$ **a** and $^4He$ **b** detection limits obtained during several tritium measurement periods extending over several weeks

noise') is 0.35 cps with an error of $\pm 0.06$ cps (standard deviation of the mean value of the data shown in Fig. 13). The error of the *single* line blank measurement (calculated from the errors of the total signal at mass $^3He$ and the error of the 'white noise') is about 0.03 cps (1 $\sigma$). This value is dominated by the error of the 'white noise' (0.022 cps) and can only be improved by a drastic reduction of the level of scattered ions. On the basis of these considerations the $^3He$ detection limit (sensitivity: 1 cps/$10^{-15}$ ccSTP $^3He$) is calculated to $\approx 6 \cdot 10^{-17}$ ccSTP (2 $\sigma$) corresponding to $\approx 0.002$ TU for a sample size of 500 $cm^3$ and a $^3He$ ingrowth time of 6 months. In reality, the tritium detection limit is higher due to additional uncertainties introduced mainly by correction for different blank contributions

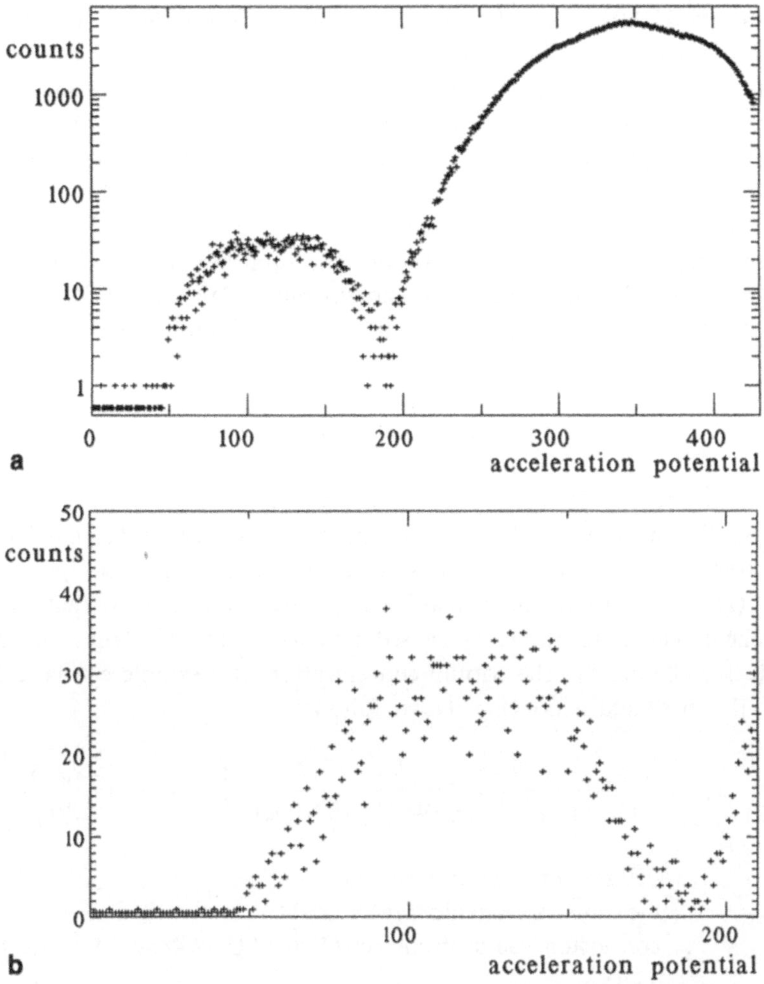


**Fig. 15. a** Scan over mass 3 in a logarithmic scale demonstrating the good resolution of $^3$He. The integration time was 2 sec using $1.4 \cdot 10^{-14}$ ccSTP $^3$He. **b** Same scan over $^3$He plotted on a linear scale. It is evident that even for low countrates the plateau is sufficiently broad to allow $^3$He measurement with good precision

(see below). The $^3$He detection limits obtained with our system ($1 \cdot 10^{-16}$ ccSTP ($2\sigma$; Fig. 14)) are close to the theoretical value for our system (0.03 cps$\equiv$ $6 \cdot 10^{-17}$ ccSTP).

A potential factor limiting the detection limit is contribution of HD/$H_3^+$ to the mass of $^3$He. In our routine procedure this contribution is included in the line blank which is subtracted from the measured signal. On the basis of scans over mass 3 with a sample of $\approx 1.4 \cdot 10^{-14}$ ccSTP $^3$He in the mass spectrometer it is evident that the $^3$He peak is well resolved from the HD/$H_3^+$ peak, consequently the $HD/H_3^+$ contribution to the $^3He$ signal should be very small (Fig. 15). From mass spectrometer blank measurements (no sample in the mass spectrometer) the $HD/H_3^+$ contribution is estimated to $\approx 0.02$ cps (pointing to an abundance sensitivity of better than $10^{-5}$; $HD/H_3^+$ count rate $\approx 3000$ cps). This count rate is equivalent to a $^3He$ amount of $2 \cdot 10^{-17}$ ccSTP and variations of this amount have practically no influence on the detection limit.

Memory effects due to varying gas loads from different samples were checked by admitting samples with high $^3He$ concentrations (50 to 100 cps) into the mass spectrometer followed by line blank measurements. We did not find any significant memory effects in the range of interest for tritium measurements.

#### 4.5.4 *Data Evaluation*

The tritiogenic $^3He$ concentration of a water sample is obtained by subtraction of all non-tritiogenic $^3He$ components from the measured signal. The excess $^4He$ (typicall $< 1.5 \cdot 10^{-9}$ ccSTP) is used to monitor these components. The $^3He/^4He$ ratios of the single blank components are close to that of the atmosphere. Therefore, the error caused by corrections for blanks are relatively small (see chapter 5). The tritium concentration of a sample can be calculated from the measured tritiogenic $^3He$ as follows:

$$c_{trit} = \frac{^3He_{trit}}{C} \cdot \frac{e^{\lambda(t_s - t_e)}}{1 - e^{-\lambda(t_e - t_m)}} \cdot \frac{1}{W \cdot [1 - (S/1000)]} \cdot \left[ 1 - (a-1) \cdot \frac{W_0 - W}{W_0} \right]$$

$c_{trit}$ ... tritium concentration in TU
$^3He_{trit}$ ... measured tritiogenic $^3He$ in ccSTP
C ... conversion factor from ccSTP to TU ($2.4889 \cdot 10^{-15}$ [(ccSTP/g)/TU])
$\lambda$ ... reciprocal of mean lifetime of tritium (17.93a, CLARKE et al., 1976)
$t_s$, $t_e$, $t_m$ ... dates of sampling, extraction and measurement, respectively
$W_0$, W ... weights of the sample before and after the extraction, respectively
S ... salinity in ‰
$a$ ... correction for the $^3H/H$ fractionation due to loss of (distilled) water during gas extraction. $a = 1.15$ (ratio of tritium concentration in the liquid phase to tritium concentration in the water vapor (CLARKE et al., 1976).

## 5 Blanks

There are many potential sources of tritium and/or $^3He$ blanks in all individual steps of the measurement procedure. These blanks have to be subtracted from the measured signal before it is converted to final data. To do the correction properly, the blank components have to be separated and the effect of the correction on the precision of the measurement has to be estimated.

## 5.1 Helium Measurement

### *5.1.1 Sampling and Sample Storage*

The first potential to influence the sample is introduction of air into the head space of the Niskin sampler. Oceanographic samples are supersaturated in $^4$He by typically 3% to 15% and in $^3$He by up to 50% or more. TAKAHASHI et al. (1988) estimated that during the first 10 minutes after introduction of air into the Niskin sampler $\approx 0.2\%$ of dissolved helium is lost to the head space, if a 10 l Niskin sampler is used, less than 50% of the water is drawn from the sampler and the filling rate is $<0.5$ l/min. Therefore, under normal conditions, $^3$He loss to the head space of the Niskin samplers should not be a serious problem, and the errors introduced by this process should be on the same order of magnitude as the measurement precision.

A more serious problem is the inclusion of small air bubbles during the filling procedure of the sample containers. From our experience the quality of helium sample sets is highly variable between different cruises, even in cases where the samples were taken by the same person. Here additional work is necessary to define the possible sources of scatter.

The helium concentration of the water samples can also be changed during storage in the copper tubes via exchange across the pinch-off clamps. We checked the pinch-off clamps developed at Heidelberg for leak rates and found values of about $10^{-9}$ (ccSTP $^4$He)/(sec·atm) for dry clamps applying pure helium. This is less than $5\cdot10^{-15}$ ccSTP $^4$He/sec for atmospheric helium concentration. During closure of the copper tube at sea, the roughness of the copper is smoothed out by a thin film of water reducing the leak rate further to values of $\approx 10^{-19}$ (ccSTP $^4$He)/(sec·atm air). This low leak rates are equivalent to an e-folding time on the order of $10^5$ years for helium exchange with the atmosphere.

If the water sample contains tritium, $^3$He is produced by tritium decay during storage. For example, a water sample with a tritium concentration of 5 TU produces a shift of the $^3$He/$^4$He ratio on the order of 2.5% in a storage time of 2 years. Therefore, we routinely correct for tritium decay during storage of the samples. The uncertainty introduced by this correction depends on the accuracy of the tritium measurement. If we use an accuracy of about $\pm 5\%$ for the tritium concentration in our example the $^3$He/$^4$He correction leads to an additional error of $\approx \pm 0.1\%$.

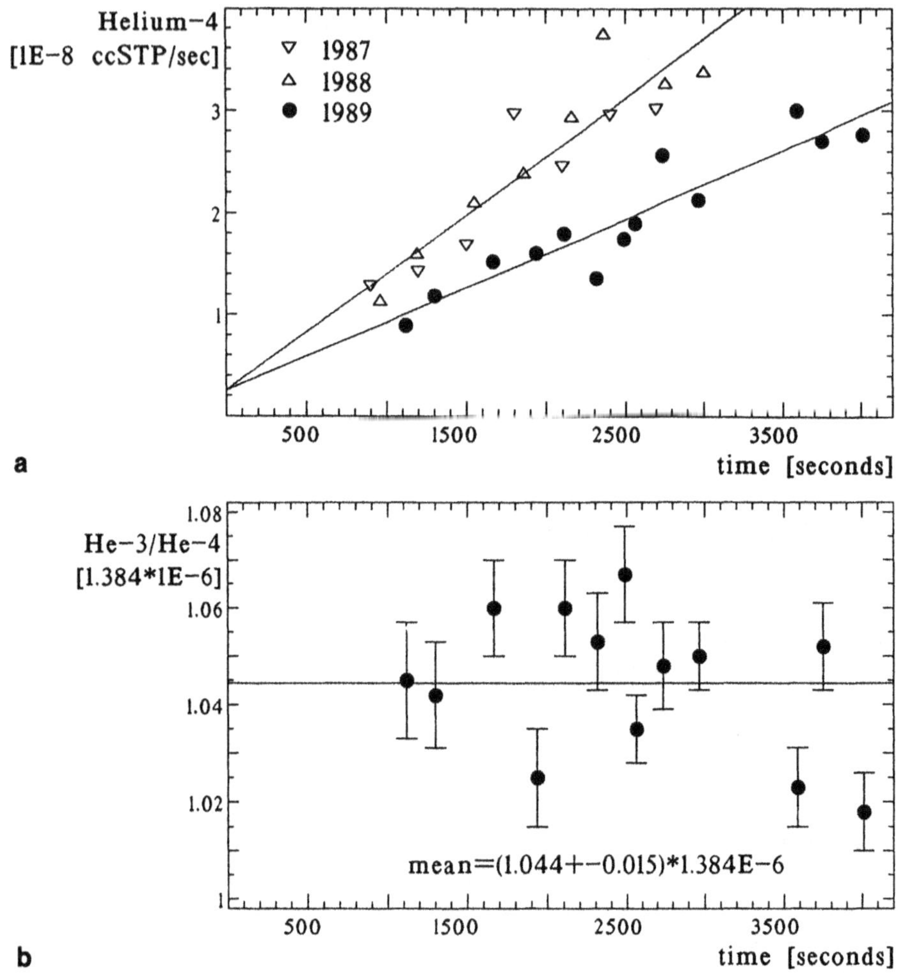


**Fig. 16. a** $^4$He concentration accumulated from permeation through 7 O-rings as a function of time for 3 different times. The different symbols indicate different O-ring types and/or different O-ring age: inverse triangle: 'old' type (used for some weeks); triangle: 'new' type (new); dots: 'new' type (used for some weeks). **b** $^3$He/$^4$He ratio of the helium permeated through the O-rings used in the 1989 experiment shown in **Fig. 16a**

### *5.1.2 Sample Extraction*

A considerable blank is caused by the 7 O-rings used in the extraction system in each of the eight extraction units. This blank was determined in the following way: About 10 cm$^3$ of water was degassed and the helium extracted from the water was pumped away. After this step a procedure identical to a normal extraction was started and the helium released by the O-rings was accumulated in the glass ampoule with different accumulation times for the single units. The measurement of the ampoules yields accumulated helium amounts increasing linearly with time (Fig. 16a). From the slopes of the fitted lines (we performed this experiment three times), the $^4$He release is estimated to about $1.6 \cdot 10^{-12}$ ccSTP/sec to $9.3 \cdot 10^{-13}$ ccSTP/sec per O-ring. The relatively high $^4$He concentrations obtained at the zero intercept of the fit curves ($\approx 2.5 \cdot 10^{-9}$ ccSTP $^4$He) are due to incomplete degassing of the water, helium release from the glass during flame sealing and storage of the glass ampoules prior to measurement. The amount of helium accumulated from the 7 O-rings during the extraction time of 30 min is $(1.21 \pm 0.14) \cdot 10^{-8}$ ccSTP which is about 0.6% of the typical helium concentration of a 40 g water sample. A routine correction for this blank is performed with an accuracy of about $\pm 0.07$%.

The $^3$He/$^4$He ratio is also influenced by the extraction blank. It is lowered in samples with $^3$He/$^4$He ratios higher than that of the helium component released by the O-rings ($(1.044 \pm 0.015) \cdot R_{air}$; Fig. 16b; with $R_{air} = 1.384 \cdot 10^{-6}$; CLARKE et al., 1976) and increased for samples with values below the O-ring $^3$He/$^4$He ratio. Taking a typical $^4$He O-ring blank and the $^3$He/$^4$He ratio of $1.044 \cdot R_{air}$, the total $^3$He release during extraction is estimated to about $(1.75 \pm 0.21) \cdot 10^{-14}$ ccSTP. The correction of this blank does not increase the error of the $^3$He/$^4$He measurement significantly.

During flame sealing of the glass ampoule the glass capillary is heated to about 1000 °C and the helium permeation rate of the glass is increased considerably. The amount of helium released during flame sealing can be estimated from the helium inventory of the heated glass volume. An upper limit of $7 \cdot 10^{-10}$ ccSTP $^4$He is obtained using a volume of the glass capillary of $\approx 0.15$ cm$^3$ and a solubility of helium in glass of about $10^{-3}$ ccSTP/(cm$^3 \cdot$atm air) (see below). This value was checked by flame sealing ampoules on-line in the inlet system of the mass spectrometer and measuring the concentration of the released helium. We obtained values ranging from $1 \cdot 10^{-10}$ to $2.5 \cdot 10^{-10}$ ccSTP $^4$He without significant isotopic fractionation, in reasonable agreement with the theoretical upper limit. In practice, the contribution of the seal-off blank to the precision of the $^3$He/$^4$He measurement is negligible.

## 5.2 Tritium Measurement

### *5.2.1 Tritium Blanks*

In principle, tritium can be added to the water samples during transfer into the sample containers and during storage of the samples. If care is taken that no water with high tritium concentrations is added to the water samples during filling of the glass bottles contamination can only occur by water vapor exchange. Under normal circumstances, i.e. if the atmospheric tritium concentration is not elevated extremely by tritium tagged watches or tritium used for tracer experiments, tritium concentrations of the water samples should not be increased significantly by water vapor exchange during the relatively short period of sampling. For storage, we use 1 l glass bottles which are sealed by a low density polyethylene cap pressed onto the glass. On the basis of water vapor exchange rates through these caps (WEISS et al., 1976) and in atmospheric tritium concentration of about 20 TU, the tritium concentration is increased by less than 0.001 TU per year. Therefore, sampling and storage of the water should not be problematic with respect to the tritium blank.

A source of tritium blank during storage might be cosmic ray induced in situ production of tritium in the water samples. The production rate should be on the order of $\leq 0.01$ TU/year, and experiments to determine this effect quantitatively are presently being performed in the laboratory of W. J. JENKINS at the Woods Hole Oceanographic Institution (JENKINS, 1989).

### *5.2.2 $^3$He Blanks*

$^3$He blanks in tritium measurement via $^3$He ingrowth can arise from incomplete degassing of the water sample before storage for $^3$He ingrowth, from diffusion of helium from the glass bulb used for storage into the water and permeation of helium through the O-rings used to attach the storage bulbs to the all-metal inlet system.

Principally the extraction blank and the blank originating from diffusion out of the glass bulb cannot be separated. However, measurement of the water directly after extraction holds the glass blank so low that the extraction blank (including the O-ring blank and the blank from flame sealing) can be estimated reasonably well if the O-ring blank and the flame sealing blank components, which both can be measured directly, are substracted. The O-ring blank was measured directly after measurement of tritium samples (wet O-rings). We obtained a mean value of about $10^{-12}$ ccSTP $^4$He/sec with a $^3$He/$^4$He ratio of about $1.05 \cdot R_{air}$ (Fig. 17). This leads to a total O-ring blank of $2.5 \cdot 10^{-10}$ ccSTP $^4$He during the sample inlet time (250 sec). After correction for this blank and subtraction of the blank component due to flame sealing of $\approx 1.5 \cdot 10^{-10}$ ccSTP $^4$He (see above), the extraction blank is estimated to about $1.5 \cdot 10^{-10}$ ccSTP $^4$He (see 4.1.2).

The blank components originating from the O-ring, the flame sealing and the extraction system add up to about ($5.5 \pm 1.0 \cdot 10^{-10}$ ccSTP $^4$He). This value can be compared to the total blanks obtained from our large volume tritium measurements using pretreated (baked out under nitrogen atmosphere at about 450 °C) bulbs stored at a temperature of about $-25$ °C. The mean value of these blank measurements is $5.4 \cdot 10^{-10}$ ccSTP $^4$HE (Fig. 18), which is very close to the sum of the O-ring blank, the flame sealing blank and the extraction blank. Therefore, we conclude that diffusion from the glass into the samples is low if the glass bulbs

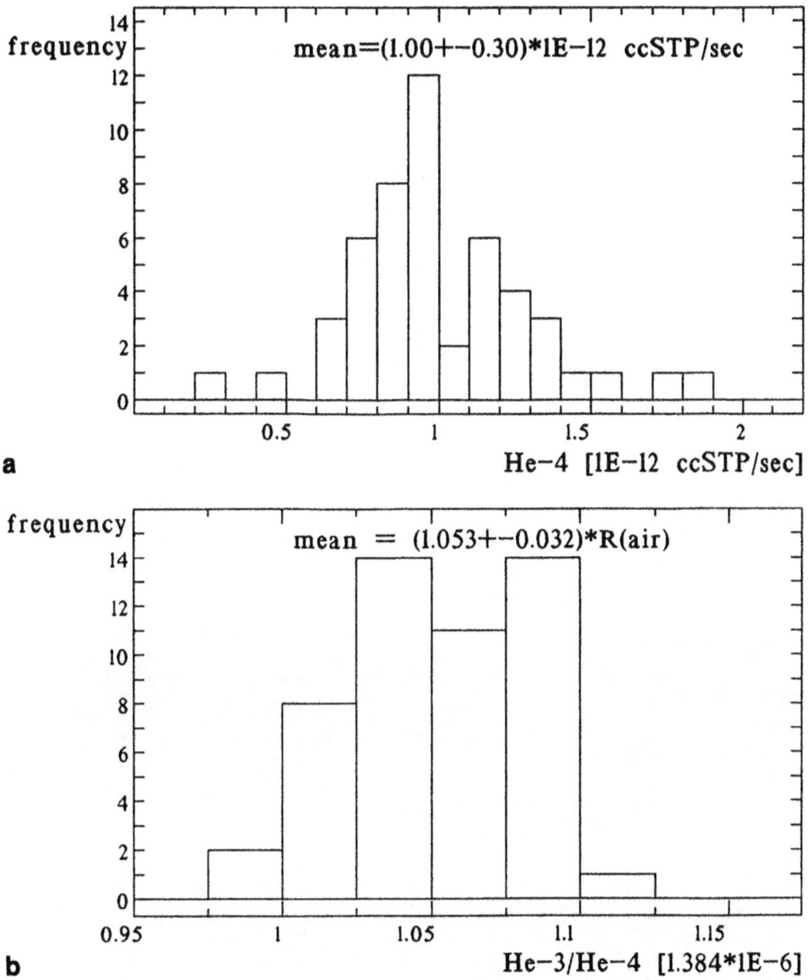


**Fig. 17.** **a** $^4$He permeation rates measured for single 'wet' O-rings (for explanation see text). **b** $^3$He/$^4$He ratio of the helium permeated through 'wet' O-rings

are pretreated and stored at low temperatures. This conclusion is supported by measurements of the $^4$He concentration accumulated in glass bulbs stored at different temperatures as a function of time (Fig. 19). From these measurements it is evident that pretreated bulbs stored at low temperatures are uncritical concerning the diffusion of helium from the glass into the water samples. However, Fig. 19 also clearly shows, that pretreatment as well as storage at low temperatures is necessary to avoid serious problems with helium blanks from the storage bulbs. From the accumulated $^4$He concentrations in glass bulbs that were not pretreated, the diffusion constant of helium in the glass can be estimated by a model simulation of the permeation out of the glass. For this purpose the solubili-

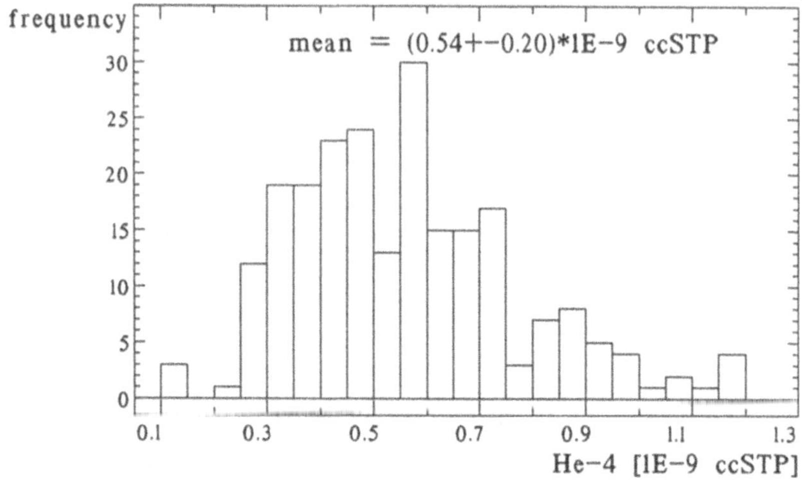


**Fig. 18.** Distribution of total blanks obtained during a tritium measurement period (226 single measurements)

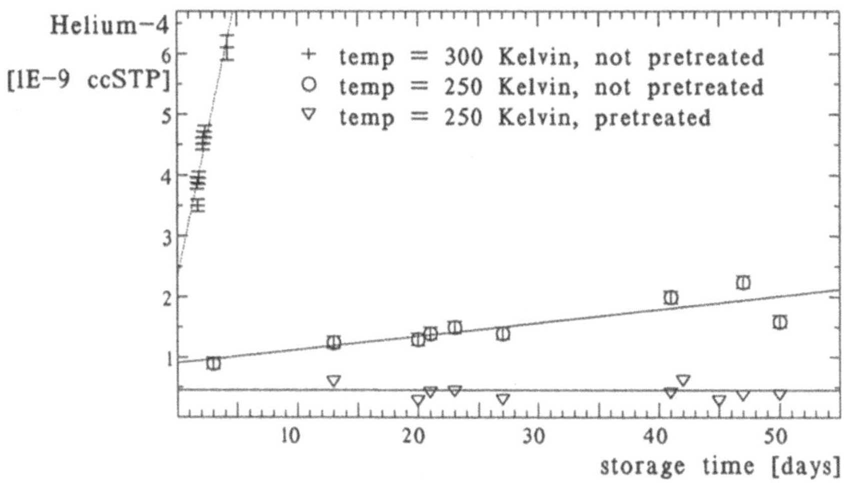


**Fig. 19.** Accumulated helium in storage glass bulbs ($A \approx 530\,cm^2$) as a function of time and storage temperature

ty of helium in glass has to be known. Taking a solubility of $\approx 8 \cdot 10^{-4}$ for GW-glass (SUCKOW, 1989), we estimated diffusion constants of $\approx 4 \cdot 10^{-6}\,cm^2/sec$ at a temperature of 450 °C and $\approx 3 \cdot 10^{-13}\,cm^2/sec$ at a temperature of $-25$ °C (Fig. 20). The measured $^3He/^4He$ ratio of the storage bulb blank component shows no significant deviation from that of atmospheric helium. Therefore, the $^3He$ blank caused by diffusion of helium out of the glass bulbs can be corrected for even in the case of a somewhat higher blank contribution of the glass.

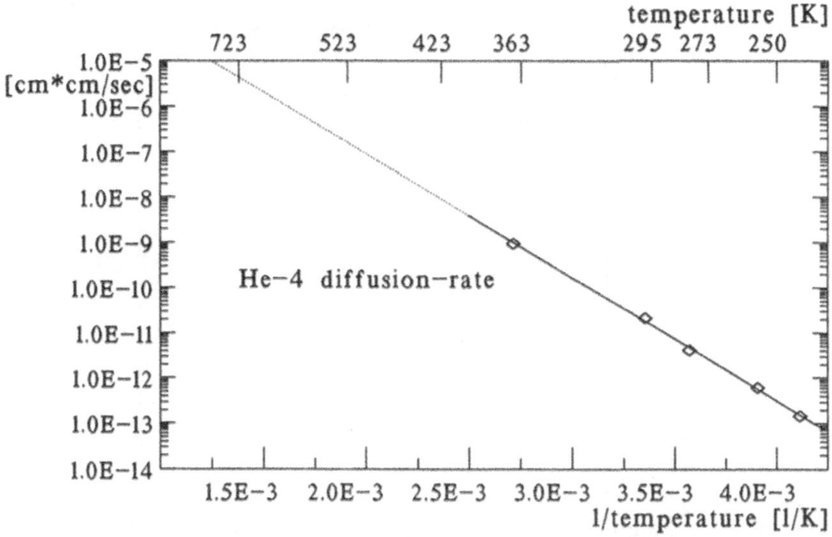


**Fig. 20.** Diffusion constants of helium in GW glass estimated from a simulation of the accumulated helium concentration in not pretreated glass bulbs (for explanation see text)

We can estimate the additional error of the tritium measurement introduced by the correction for the blank components if we take the observed $^3$He amounts of the individual blank components and the corresponding measured or estimated $^3$He/$^4$He ratios. If we assume a $^3$He/$^4$He ratio of $1.15 \cdot R_{air}$, i.e. the theoretical maximum fractionation, for the extraction and flame sealing and storage bulb components, and the measured $^3$He/$^4$He ratio for the O-ring blank ($\approx 1.05 \cdot R_{air}$), we obtain an upper limit of $\pm 1.4 \cdot 10^{-16}$ ccSTP $^3$He. This value corresponds to a tritium concentration of about ±0.004 TU for a 500 g sample stored for $^3$He ingrowth for 6 months.

## 6 'Ultra Low Level' Tritium Measurements

To demonstrate the capability of the system a tritium profile from the Weddell Sea (Antarctica) is shown in Fig. 21. These measurements clearly show that in the southern hemisphere the structure of the tritium profiles can be only resolved by measurement of tritium via $^3$He ingrowth. The good agreement of the duplicate samples is evidence for the overall reproducibility of the measurements. The non-zero values in the tritium minimum are expected from the turnover times of the water masses of the Weddell Sea and can not be interpreted as an offset in our measurements.

Measurement of samples with extremely low tritium concentrations (brine taken in ≈1000 m depth in an iron mine at Salzgitter, FRG) give evidence that

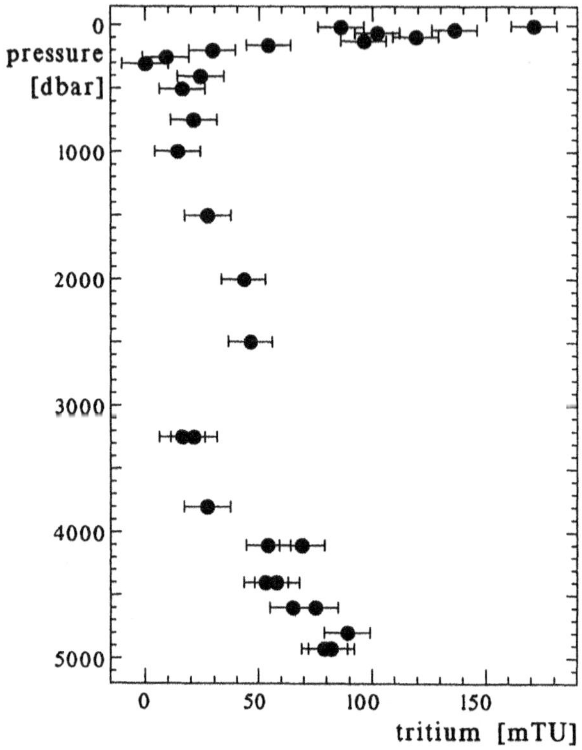


**Fig. 21.** Tritium profile from station 245 occupied in the Weddell Sea during the third expedition of the German Polar Research Vessel *Polarstern* to Antarctica (ANT III/3, January to March, 1985). Position: 66.77 °S, 21.80 °W; date of sampling: January 17, 1985

detection limits well below 0.005 TU can be obtained with our system (see Table 1). For these measurements we used a sample size of about 500 g and the samples were stored for about 4 months to allow $^3He$ ingrowth. The samples were degassed and measured routinely.

## 7 Intercalibration

For small volume mass spectrometric tritium measurement an intercalibration with the Heidelberg low-level $\beta$-counting laboratory was done. For this purpose samples taken on a station in the Arctic Ocean were measured both mass spectrometrically and using the low-level counting technique. The accuracy of the $\beta$-counting results is $\pm 5\%$ with a detection limit ($1\sigma$) of about 0.08 TU (WEISS et al., 1976). From the result (Fig. 22) it is evident that the errors of the mass spectrometric measurements are smaller than those of the $\beta$-counting results, and that within the errors the data show a good agreement. The slope of the fit curve

**Table 1.** Measurements on May 30, 1988. S 4051/1 Ort 300 and S 4051/2 Ort 300 as well as S 4053/1 Hils and S 4053/2 Hils are independent measurements of the same water samples. The size of the standard aliquot was $2.69 \cdot 10^{-15}$ ccSTP $^3He$. Errors of the tritium measurement are 1 $\sigma$ errors

| Sample name | | $^4He$ [mV] | $^3He$ [cps] | $^4He$ [$10^{-10}$ cc[a]] | $^3He$ [$10^{-16}$ cc[a]] | $^3He_{trit}$ [$10^{-17}$ cc[a]] | Tritium [TR] |
|---|---|---|---|---|---|---|---|
| A[b] | 1-0202 | 3.918 | 2.69 | | | | |
| A | 2-0304 | 7.674 | 5.34 | | | | |
| S*[c] | T420 NW-T4 | 1.316 | 1.11 | 5.74 | 8.85 | 3.98 | |
| S* | T417 NW-T1 | 1.198 | 1.03 | 5.12 | 7.99 | 3.75 | |
| A | 3-0507 | 11.500 | 3.34 | | | | |
| S[c] | 4053/1 Hils | 1.130 | 1.01 | 4.77 | 7.75 | 6.25 | 0.004 ± 0.004 |
| B[d] | 1 | 0.248 | 0.29 | | | | |
| S* | T418 NW-T2 | 1.088 | 0.89 | 4.56 | 6.41 | −4.18 | |
| A | 1-0808 | 3.924 | 2.68 | | | | |
| S | 4051/1 Ort 300 | 1.638 | 1.32 | 7.40 | 11.10 | 3.65 | 0.002 ± 0.005 |
| B | 2 | 0.243 | 0.35 | | | | |
| A | 2-0910 | 7.672 | 5.16 | | | | |
| S | 4051/2 Ort 300 | 1.262 | 1.12 | 5.45 | 8.94 | 8.71 | 0.006 ± 0.005 |
| S | AIR 0.1558 $cm^3$ | 1.676 | 1.23 | 7.60 | 1.01 | (−3.9) | |
| S | 4053/2 Hils | 1.056 | 0.92 | 4.39 | 6.79 | 1.88 | 0.001 ± 0.005 |
| A | 1-1111 | 3.922 | 2.75 | | | | |

[a] STP... standard temperature and pressure.
[b] A ... air standard.
[c] S ... sample.
S* ... samples with low tritium concentration measured directly after gas extraction.
[d] B ... blank.

Sensitivity $^4He$ = 3.752 mV/Aliquot (±0.65%). Blank $^4He$ = 0.207 mV (±22.1%).
Sensitivity $^3He$ = 2.45 cps/Aliquot (±0.86%). Blank $^3He$ = 0.30 cps (±13.0%).
Detection limit $^4He = 4.7 \cdot 10^{-11}$ ccSTP $^4He$ (2$\sigma$).
Detection limit $^3He = 8.7 \cdot 10^{-17}$ ccSTP $^3He$ (2$\sigma$).

Tritium detection for given parameters: 0.005 TR (2$\sigma$).

(0.998±0.014) is not significantly different from 1, and no significant offset at zero intercept (0.019±0.023 TU) can be observed. This is evidence that within the accuracies of the single measurement techniques the two data sets are in full agreement.

## Acknowledgements

This work would not have been possible without the substantial contribution of many persons and institutions. Wolfgang ROETHER, the former head of the

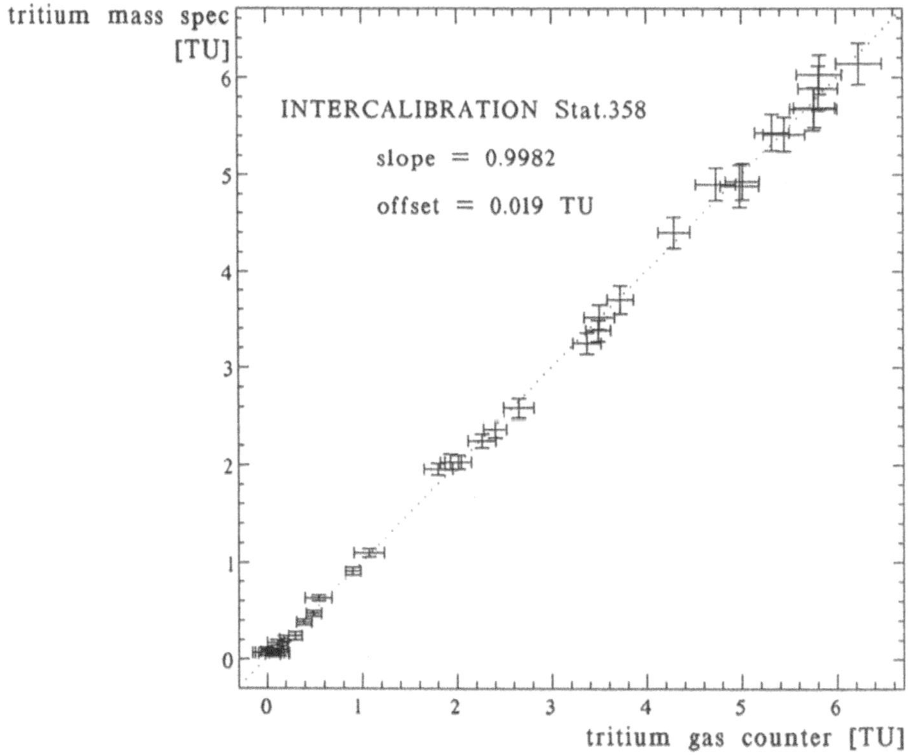


**Fig. 22.** Intercalibration between mass spectrometric tritium measurements (40 g samples) and the Heidelberg low-level $\beta$-counting laboratory. The error bars indicate the 1 $\sigma$ error

Oceanography group at our institute (now at University of Bremen) was responsible for the project between 1981 and 1987 and contributed to numerous technical and scientific aspects. Karl Heinz FISCHER designed and set up major parts of the system and was heading the helium group until 1984. Wolfgang WEISS started up the project and contributed to a variety of aspects in the initial phase. The project greatly profited from the generous support given by the Max-Planck-Institut für Kernphysik, especially from Till KIRSTEN and Dieter DÖRFLINGER. This support made it much easier for us to enter the field of helium isotope measurement and learned us the ultra-high vacuum technique. Gerhard FUCHS helped to develop the inlet system and the electronics needed for automatic measurement. Karl Otto MÜNNICH encouraged us during the whole project and gave valuable comments to numerous aspects of the system design. The mechanical workshop of the institute headed by Reiner FLETTERER helped to design and manufactured many parts of the system. Christian SONNTAG provided the groundwater used to check the detection limit of the system and Gerhard BADER performed the low-level $\beta$-counting measurements used for intercalibration. This project was funded by the Deutsche Forschungsgemeinschaft.

## References

ATAKAN Y, ROETHER W, MÜNNICH KO, MATTHESS G (1974) The Sandhausen shallow-groundwater tritium experiment. In: Isotope Techniques in Groundwater Hydrology 1974, Vol. 1, pp 21–43, IAEA-SM-182/2, IAEA, Vienna

BEGEMANN F, LIBBY WF (1957) Continental water balance, ground water inventory and storage times, surface ocean mixing rates and world-wide water circulation patterns from cosmic-ray and bomb tritium. Geochim Cosmochim Acta 12:277–296

CLARKE WB, BEG MA, CRAIG H (1969) Excess $^3He$ in the sea: Evidence for terrestrial primordial helium. Earth Planet Sci Lett 6:515–522

CLARKE WB, JENKINS WJ, TOP Z (1976) Determination of tritium by mass spectrometric measurement of $^3He$. International Journal of Applied Radiation and Isotopes 27:217–225

JENKINS WJ (1989) pers. communication

JENKINS WJ, CLARKE CB (1976) The distribution of $^3He$ in the western Atlantik Ocean. Deep-Sea Res 23:481–494

JENKINS WJ, LOTT DE, PRATT MW, BOUDREAU RD (1985) Anthropogenic tritium in South Atlantic bottom water. Nature 305:45–46

LYNCH MFC, KAY DJ (1981) Performance of a mass spectrometer for determining low tritium levels from $^3He/^4He$ measurements. In: Methods of Low-Level Counting and Spectrometry, pp 511–523, IAEA-SM-252/47, IAEA, Vienna

MÜNNICH KO, ROETHER W (1967) Transfer of bomb $^{14}C$ and tritium from the atmosphere to the oceans: internal mixing of the ocean on the basis of tritium and $^{14}C$ profiles. In: Radioactive dating and Methods of Low Counting, pp 93–103, IAEA, Vienna

ÖSTLUND HG, BRESCHER R (1982) GEOSECS Tritium. Tritium Laboratory Data Report No. 12, Rosenstiel School of Marine and Atmospheric Science, University of Miami, pp 205

REYNOLDS JH, JEFFERY PM, MCCRORY GA, VARGA PM (1978) Improved charcoal trap for rare gas mass spectrometry. Rev Sci Instruments 49:547–548

SCHLOSSER P (1986) Helium: a new tracer in Antarctic oceanography. Nature 321:233–235

SCHLOSSER P, STUTE M, DÖRR H, SONNTAG C, MÜNNICH KO (1988) Tritium/$^3He$ dating of shallow groundwater. Earth Planet Sci Lett 89:363–374

SUCKOW A (1989) Diffusion und Löslichkeit von Helium in verschiedenen Glassorten. Diplomarbeit, in Vorbereitung

TAKAHASHI T, WILLIAMS RT, CHIPMAN DW, CEMBER R (1988) Alteration of the concentrations of dissolved gases in Niskin Bottles during sampling operations. SAVE Technical Report, unpublished document

TORGERSEN T, CLARKE WB, JENKINS WJ (1979) The tritium/helium-3 method in hydrology. IAEA-SM-228/49:917–930, IAEA, Vienna

WEISS W, ROETHER W, BADER G (1976) Determination of blanks in low-level tritium measurement. International Journal of Applied Radiation and Isotopes 27:217–225

# Inhalt
## Jahrgang 1989

## Sitzungsberichte der Heidelberger Akademie der Wissenschaften
## Mathematisch-naturwissenschaftliche Klasse

*Die Jahrgänge bis 1921 einschließlich erschienen im Verlag von Carl Winter, Universitätsbuchhandlung in Heidelberg, die Jahrgänge 1922–1933 im Verlag Walter de Gruyter & Co. in Berlin, die Jahrgänge 1934–1944 bei der Weißschen Universitätsbuchhandlung in Heidelberg. 1945, 1946 und 1947 sind keine Sitzungsberichte erschienen.*

*Ab Jahrgang 1948 erscheinen die „Sitzungsberichte" im Springer-Verlag.*

**Inhalt des Jahrgangs 1984:**

1. R. Lüst. Extraterrestrische Astronomie. DM 17,–.
2. F. Leonhardt. Zu den Grundfragen der Ästhetik bei Bauwerken. DM 12,–.
3. Ch. Rüchardt. Die Bindung zwischen Kohlenstoffatomen, das Rückgrat der Organischen Chemie, und ihre Grenzen. DM 12,80.
4. J. Peiffer. Zur Neuropathologie der Nebenwirkungen nervenärztlicher Therapie. DM 18,–.
5. F. Linder. Geistige Grundlagen der chirurgischen Therapie. DM 14,–.

Medizinische Anthropologie. Herausgegeben von E. Seidler. Supplement. Geb. DM 76,–.

W.-W. Höpker. Mißbildungen. Interrelationen, Assoziationen und diagnostische Validität. Supplement. Geb. DM 74,–.

**Inhalt des Jahrgangs 1985:**

1. H. A. Staab. Zur Entstehung des Neuen in den Naturwissenschaften – dargestellt an einem Beispiel der Chemiegeschichte. DM 16,50.
2. S. Sambursky. Proklos, Präsident der platonischen Akademie, und sein Nachfolger, der Samaritaner Marinos. DM 13,–.
3. R. Haas. AIDS – Ein Virusinfekt des Immunsystems. DM 21,50.
4. F. Räbiger. Beiträge zur Strukturtheorie der Grothendieck-Räume. DM 39,50.
5. W. Kaiser. Entwicklungslinien der Breitbandkommunikation. DM 22,–.

Pathogenese. Herausgegeben von H. Schipperges. Supplement. Geb. DM 88,–.

E. Hinz. Human Helminthiases in the Philippines. Supplement. Geb. DM 98,–.

T. Cremer. Von der Zellenlehre zur Chromosomentheorie. Supplement. Geb. DM 135,–.

**Inhalt des Jahrgangs 1986:**

1. W. Doerr. Hat das Menschengeschlecht eine biologische Zukunft? DM 22,50.
2. G. Schettler. Der Stoffwechsel der Plasmalipoproteine und seine Bedeutung für die Pathogenese der Arteriosklerose. DM 38,–.
3. A. Fröhlich. Tame Representations of Local Weil Groups and of Chain Groups of Local Principal Orders. DM 55,–.
4. W. Doerr. Pathologie in Heidelberg. Stufen nach 1945. DM 14,80.

**Inhalt des Jahrgangs 1987/88:**

1. H. Schipperges. Eine „Summa Medicinae" bei Avicenna. Zur Krankheitslehre und Heilkunde des Ibn Sīnā (980–1037). DM 34,80.
2. H. Elsässer. Aktive Galaxien. DM 32,–.
3. W. Rauh. Tropische Hochgebirgspflanzen. Geb. DM 98,–.

GPSR Compliance

*The European Union's (EU) General Product Safety Regulation (GPSR) is a set of rules that requires consumer products to be safe and our obligations to ensure this.*

*If you have any concerns about our products, you can contact us on ProductSafety@springernature.com*

In case Publisher is established outside the EU, the EU authorized representative is:

Springer Nature Customer Service Center GmbH
Europaplatz 3
69115 Heidelberg, Germany

**Batch number: 10704763**

Printed by Printforce, the Netherlands